# Field-Effect Transistors - Fundamentals, Technologies, and Future Innovations

*Edited by Kenan Cicek*

Published in London, United Kingdom

Field-Effect Transistors - Fundamentals, Technologies, and Future Innovations
http://dx.doi.org/10.5772/intechopen.1009040
Edited by Kenan Cicek

Contributors
Ahmed Yusupov, Atabek Atamuratov, Debasish Mohanta, Jean-Marc Dienot, Kenan Cicek, Mahdieh Nayeri, Makhkam Khalilloev, Mandela Srikanth, Maryam Nayeri, Prasanna Kumar Sahu, Revanth Bokka, Sruti Suvadarsini Singh, Uday Kumar Dosanapudi



First published in London, United Kingdom, 2026 by IntechOpen
IntechOpen is the global imprint of INTECHOPEN LIMITED, registered in England and Wales, registration number: 11086078, 167-169 Great Portland Street, London, W1W 5PF, United Kingdom

For EU product safety concerns: IN TECH d.o.o., Prolaz Marije Krucifikse Kozulić 3, 51000 Rijeka, Croatia, info@intechopen.com or visit our website at intechopen.com.

British Library Cataloguing-in-Publication Data
A catalogue record for this book is available from the British Library

Field-Effect Transistors - Fundamentals, Technologies, and Future Innovations
Edited by Kenan Cicek
p. cm.
Print ISBN 978-1-83635-007-1
Online ISBN 978-1-83635-006-4
eBook (PDF) ISBN 978-1-83635-008-8

# Meet the editor

Kenan Cicek, Ph.D., is an Associate Professor of Electrical and Electronics Engineering at Iğdır University, Türkiye. He received his Ph.D. from the University of Bristol, UK, in 2016, specializing in photonic integrated circuits, optical sensors, plasmonics, and microwave photonics. He has extensive experience in simulation, fabrication, and interdisciplinary research. His work has been widely published in high-impact journals, and he has contributed to projects funded by prestigious scientific organizations. Dr. Cicek was selected for the Republic of Türkiye Ministry of National Education's 2012 YLSY Scholarship Program and was awarded a Fulbright Fellowship at the Massachusetts Institute of Technology, USA, in 2021.

# Contents

# Preface

Field-effect transistors have been one of the main driving forces behind technological evolution and have remained a central focus of scientific research for decades. From the beginning of silicon-based devices to nanoscale CMOS architectures, FET transistors have continued to shape the digital world. Throughout this journey, numerous challenges have been addressed to develop more efficient, reliable, and scalable devices for future technologies. The book, *Field-Effect Transistors – Fundamentals, Technologies, and Future Innovations*, brings together a set of chapters that explore the innovations and challenges associated with FET technologies, written by leading scientists in the field.

Each chapter of this book focuses on a distinct topic and highlights recent breakthroughs along with the associated challenges they bring. The book begins with the *Introductory Chapter: Field-Effect Transistors – Shaping the Digital Future*. This first chapter provides a brief overview of the historical development of FET transistors, explains their scientific role in current technology, and discusses their future potential in the digital world.

The book continues with the chapter: *A Review on Analytical and Electrical Modeling of the MOSFET Transistor*, which outlines a comprehensive review of MOSFET devices analysed both electrically and mathematically for circuit design applications. The chapter provides predictive simulation results addressing challenges that range from basic device-level to PCB-level power MOSFET transistors.

Self-heating is a serious issue in FET transistor circuits, especially as the miniaturisation continues. The chapter, *Self-Heating Effects in Nanoscale MOSFETs*, sheds light on this topic, focusing on FinFET transistors. It investigates how oxide materials and channel geometries affect the thermal behaviour of these transistors. It provides guidance on mitigating self-heating issues in next-generation FET technologies.

The chapter, *Challenges and Benefits of Multi-Value Digital Circuit Design with GNRFET and CNTFET Technology*, presents alternative materials to silicon as silicon approaches its physical limits. The chapter discusses GNRFETs and CNTFETs transistors in the context of ternary logic circuits. It also addresses the challenges ranging from fabrication to integration, as well as the future prospects of these transistors in digital electronics and next-generation computer systems.

The chapter, *Performance Assessment of GaSb-Si Vertical TFET with Triple Metal Gate for Low Power Application*, analyzes the performance of the GaSb-Si triple-metal-gate vertical tunnel field-effect transistor (TMG-VTFET). It presents a performance study of the power consumption and switching behaviour of these transistors, focusing on the effect of using a narrow-bandgap material in the source region.

In the chapter, *Randomized Algorithm-Based Optimization of Next-Generation Field-Effect Transistors: Design, Variability, and Performance Innovations*, a general framework for randomized algorithm-based optimization of next-generation FETs is introduced to enhance innovation performance. Various techniques, including genetic algorithms, Monte Carlo methods, and machine learning, are used to assess parameters such as geometry, doping, and material selection to improve device performance.

As the chapters demonstrate, all contributing authors dedicated considerable effort to creating this book. We sincerely appreciate their valuable contributions, which have made this work possible. The results of this collective endeavour will serve as a strong and reliable resource for the literature, benefiting not only researchers but also students at all levels.

**Kenan Cicek**
Department of Electrical and Electronics Engineering,
Igdir University,
Igdir, Türkiye

**Chapter 1**

# Introductory Chapter: Field-Effect Transistors – Shaping the Digital Future

*Kenan Cicek*

## 1. What is transistor

Transistors are the key component in electronic technology and form the foundation of nearly all electronic devices used in our daily life by controlling the flow of current or voltage. Transistors are first developed in 1956 at Bell Laboratories by three scientist William Shockley, Walter Brattain, and John Bardeen [1]. The transistors have been designed as an alternative to the vacuum tubes that had been extensively used at the time and initiated a fundamental transformation in the field of electronics. Owing to transistors' compact size, energy-efficient operation, and improved operational reliability over, electronic systems rapidly became more efficient and compact. With the introduction of the MOSFET (metal-oxide-semiconductor field-effect transistor) in 1959 by Dawon Kahng and Mohamed Atalla at Bell Labs, computers that once occupied entire rooms have become more portable and even handheld [2]. As a result, millions and eventually billions of transistors can be integrated onto a single chip to realize complex circuits such as microprocessors and memory devices.

In accordance with Moore's law, the density of transistors on a chip is expected to double approximately every 2 years. As illustrated in **Figure 1**, this continuous transistor scaling has significantly improved speed, integration density, and overall performance. This progress has led to a digital world dominated by MOSFET technology, driving the evolution of computing systems from mainframes to modern artificial intelligence (AI) platforms. However, aggressive device scaling has introduced several physical and technological challenges, motivating the transition from planar MOSFETs to advanced architectures of Fin Field-Effect Transistors (FinFETs) then gate-all-around FET transistors (GAAFETs). These three architectures are the main designs that help to overcome the limitations while enabling continued performance scaling, energy efficiency, and device reliability at nanometer technology nodes.

*Field-effect transistors - Fundamentals, technologies, and future innovations* offers a broad overview of recent progress and emerging trends in field-effect transistor technologies. It covers a wide range of topics from device physics and modeling to circuit concepts and optimization approaches, reflecting the dynamic and evolving nature of today's transistor research. The chapters discuss the key challenges at both the physical and operational levels and investigate alternative materials and device architectures to enhance performance and energy efficiency beyond conventional technologies. By combining fundamental concepts with practical perspectives, this book seeks to



IntechOpen

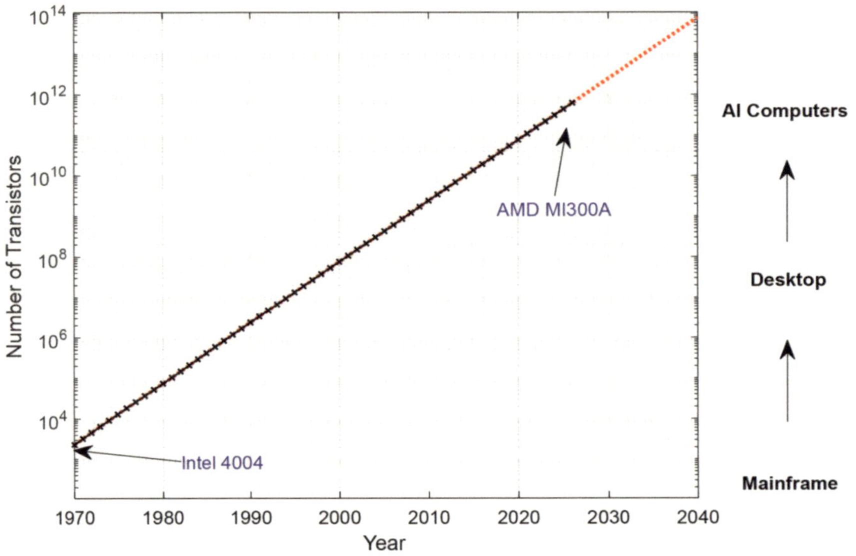


**Figure 1.**
*Scaling trends of transistor counts in computer processor chips.*

provide researchers, engineers, and graduate students with a clear understanding of the challenges and opportunities shaping the next generation of electronic devices.

## 2. Fundamentals of transistors

Transistors are one of the fundamental components of modern digital technology and are semiconductor devices that operate to amplify or switch electrical signals. As illustrated in the diagram in **Figure 2**, transistors are structurally classified into two main categories of bipolar junction transistors (BJTs) and field-effect transistors (FETs). In BJTs, both electrons and holes contribute to electrical conduction, whereas in FETs, conduction is carried out by only one type of charge carrier, either electrons or holes.

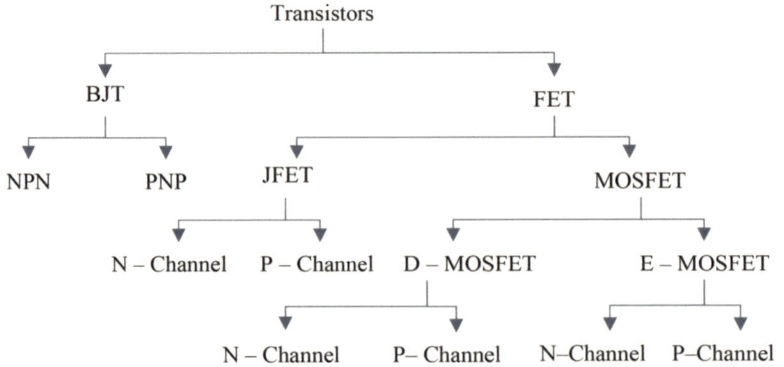


**Figure 2.**
*Structural clasification of transistors.*

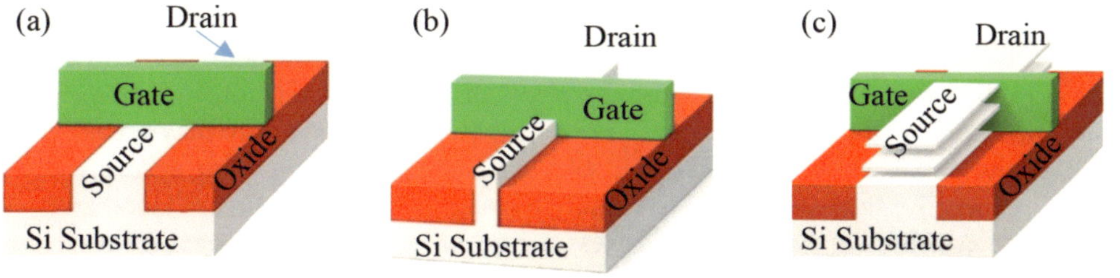


**Figure 3.**
*Transistor architectures of (a) planar FET, (b) FinFET, (c) GAAFET.*

While both BJT and FET families have played important roles in the evolution of semiconductor technology, modern digital integrated circuits are predominantly based on FET devices, particularly MOSFETs. Like other FETs, MOSFETs control the channel current through an electric field applied at the gate terminal, operating with essentially two-dimensional (2D) electrostatic control, as illustrated in **Figure 3(a)**. As CMOS technology has continued to scale into the deep nanometer regime, conventional planar MOSFETs have encountered fundamental physical limitations, including short-channel effects, gate-induced drain leakage, and electrostatic control over the channel [3]. To address these challenges and sustain performance scaling, the industry transitioned from 2D to three-dimensional (3D) device architectures by raising the channel above the substrate, leading to the introduction of FinFETs, as shown in **Figure 3(b)** [4–6]. However, as device scaling progressed, fin height limitations emerged related to carrier density, and the electrostatic control of the gate. To overcome these issues, multiple fins are employed instead of a single fin, which enhanced the drive current and overall transistor performance [7, 8]. Nevertheless, further increases in the number of fins eventually reaches practical and electrical limits, restricting additional improvements in drive current without introducing new challenges. Therefore, a new device architecture is developed in which the channel is fully surrounded by the gate, enabling much stronger electrostatic control over the channel. In this so-called GAA approach, multiple horizontally stacked channel sheets are arranged such that the gate wraps entirely around each sheet, as shown in **Figure 3(c)**. This architecture enables reduced leakage current, improved threshold voltage control, and enhanced scalability compared to FinFETs [9].

As the scaling of FETs continues to progress, the physical limits remain, and ongoing challenges are being tackled through the development of new device architectures and the search for alternative materials.

## 3. Future directions in FET technology

Looking ahead, remarkable progress in field-effect transistor research has been achieved for the last few decades. However, sustaining continuous improvements in the performance and energy efficiency has become increasingly challenging as conventional CMOS scaling approaches constantly reach their limits. Since the traditional planar MOSFETs are no longer sufficient to meet these demands on their own, attention has shifted toward new device architectures such as FinFETs and GAA transistors. These structures offer superior electrostatic control of the channel, which effectively reduce the short-channel effects, leakage currents, and power consumption. At the same time, intensive efforts are being made to explore alternative channel materials beyond silicon, including wide-bandgap semiconductors such as gallium nitride (GaN) [10] and silicon carbide (SiC) [11], as well as graphene [12] and other high-mobility materials, to overcome the physical and thermal limitations. Beyond faster and smaller logic devices, FET research is also expanding into new areas such as low-power electronics, sensors, and brain-inspired computing systems. Consequently, FETs are expected to remain a cornerstone of the future electronic systems, continuously evolving to address the growing demands of next-generation technologies.

## Acknowledgements

The author acknowledges the use of AI for language polishing of the manuscript.

## Author details

Kenan Cicek
Department of Electrics and Electronics Engineering, Igdir University, Turkey

*Address all correspondence to: eexkc@my.bristol.ac.uk

## References

[1] Hoddeson L. John Bardeen: The only person to win two Nobel prizes in physics. Physics Education. 2011;**46**:661

[2] Dawon K. Electric Field Controlled Semiconductor Device. Vol. 3. Washington, DC: U.S. Patent and Trademark Office; 1963. pp. 102-230

[3] Karimi K, Fardoost A, Javanmard M. Comprehensive review of FinFET technology: History, structure, challenges, innovations, and emerging sensing applications. Micromachines (Basel). 2024;**15**:1187

[4] Maurya RK, Bhowmick B. Review of FinFET devices and perspective on circuit design challenges. Silicon. 2022;**14**:5783-5791

[5] Chien ND, Thai BH, Shih CH. Thin-body effects in double-gate tunnel field-effect transistors. Journal of Applied Physics. 2024;**57**:215301

[6] Madhavi KB, Tripathi SL. Electrical characterization of highly stable 10nm triple-gate FinFET for different contacts and oxide region materials. Silicon. 2022;**14**:12281-12291

[7] Lin J. Advancement and challenges of field effect transistors based on multi-gate transistor. Journal of Physics: Conference Series. 2022;**2370**:012004

[8] Wimer S. Planar CMOS to multi-gate layout conversion for maximal fin utilization. Integration. 2014;**47**:115-122

[9] Radosavljevic M, Kavalieros J. Taking Moore's law to new heights: When transistors can't get any smaller, the only direction is up. IEEE Spectrum. 2022;**59**:32-37

[10] Yadav S, Thakur V, Kaul A, Das A, Rewari S, Nand D. Gallium nitride gate-all-around macaroni field effect transistor (GaN-GAA-MCFET): Biosensor for DNA detection. In: 2025 Devices for Integrated Circuit (DevIC 2025). Kalyani, India: IEEE; 2025. pp. 348-352

[11] Yadav S, Rewari S. Dielectric modulated 4 H-SiC source triple metal gate-all-around silicon carbide FET (DM-TMGAA-SiCFET) biosensor for label-free biomolecule detection. Microsystem Technologies. 2025;**31**:593-610

[12] Cai Q, Ye J, Jahannia B, Wang H, Patil C, Al Foysal Redoy R, et al. Comprehensive study and design of graphene transistor. Micromachines. 2024;**15**:406

Chapter 2

# A Review on Analytical and Electrical Modeling of the MOSFET Transistor

*Jean-Marc Dienot*

## Abstract

Power semiconductor MOSFET and other MOS-controlled devices benefit from material and technology improvements to respond to high-level power features, high voltage, high current density, short switching times, and thermal constants, which optimize energy efficiency. These enhanced characteristics induce more electromagnetic noises and temperature-management constraints for the deployment of this technology. We describe synthetically modeling theory and technic, from basic-to-advanced, to derive predictive simulations for the power MOSFET challenging issues. 1. Analytical and electrical circuit model of the MOSFET elementary cell at semiconductor level, time-domain simulation. 2. Distributed and propagative model, including device packaging and power-printed circuit board (PPCB) PEEC and 3D model levels, signal integrity simulation, common mode emission simulation, and radiated field simulation. 3. Electro-thermal model with thermal propagative network model coupled with electrical model at circuit level, time multi-domain simulation. Case studies on Power PCB with MOSFET Si et SiC illustrate modeling procedures.

**Keywords:** power MOSFET, partial element equivalent circuit, silicon carbide (SiC), numerical model, electrothermal model

## 1. Introduction

Since the 1970s, the metal oxide semiconductor field-effect transistor (MOSFET) has become one of the major active devices, both for digital and power electronic applications. As denoted a "unipolar" transistor, it comprises three semi-conductor regions, the drain, the source and the gate, with different carrier doping profiles [1]. The general principle of MOSFET consists of modulating the current flow $I_{Drain\text{-}Source}$ at the output with a voltage input control ($V_{Gate\text{-}Source}$). This creates an electric field that changes the equivalent resistance $R_{Drain\text{-}Source}$ in the semiconductor area of the output [2]. With appreciable on-state current carrying capability and off-state blocking capability, they replace main power BJTs in many power operations since the 1980s, because of their low and medium power features [3]. Since the 1990s, there were main steps in power electronic technology, first with the apparition of the

 IntechOpen

insulated gate bipolar transistor (I.G.B.T.) and now with the technological maturity of wide band gap (WBG) semiconductor material, as silicon carbide (SiC) and gallium nitride (GaN) [4]. So, all MOSFET-based transistors and topology are becoming the standards for a full wide range of applications from low to high power capabilities. Our propose is to focus on the global modeling ways for this large MOSFET family of active devices. We will remind and review synthetically all the modeling steps and thinkings, from physical semiconductor level to "external" behavioral boards and modules, so that engineers, searchers, users can continuously adapt their understanding to their actual need or to future improved chips and circuits.

## 2. MOSFET semiconductor principles

### 2.1 Elementary MOSFET cell

The elementary initial topology of the MOSFET transistor is a three-layer structure of semiconductor, with alternating p-type and n-type doping [5]. The zone with the single type of doping (n or p), between the source and the drain, is called the channel, illustrated at **Figure 1**. It is "under" the gate layers ($SiO^2$ and Al), and its doping type gives the polarity procedure to open or close the transistor in switching modes. For common and convenient use with a positive input voltage $V_{Gate\text{-}Source}$ ($V_{GS}$), the substrate and the channel zone would be initially of the p-type doping, so the transistor is in the "OFF" state.

When a positive voltage is applied on the gate, negative carriers will start to accumulate between the n-zones of drain and source. When $V_{GS}$ will be equal or higher than a threshold voltage denoted $V_{TH}$, n-type carriers will be sufficient to offer a conduction path with minimum of resistance between the drain and the source (**Figure 2**). The transistor is in the "ON" state, with an n-channel conduction mode. The thickness of the gate oxide, the width of the channel, and other parasitic electrical connections between semiconductors, metal, and dielectric material will determine the amount of current flow that can be available at the output. These first physical and technological quantities will give the first essential analytical parameters for starting electrical modeling of the MOSFET:

$t_{ox}$: the thickness of the oxide

$C_{ox}$: the equivalent capacitance of the metal-oxide-semiconductor layers of the gate

$V_{TH}$: the threshold voltage on the gate, that indicates the start of the conduction mode,

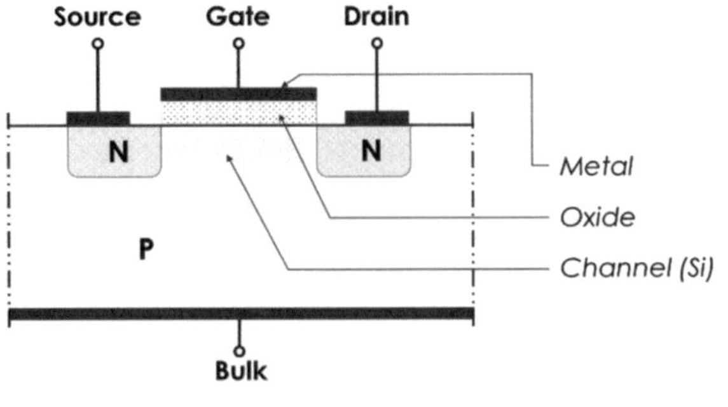


**Figure 1.**
*Physical cross-section view of the Metal-Oxide-Semiconductor Field Effect Transistor (MOSFET).*

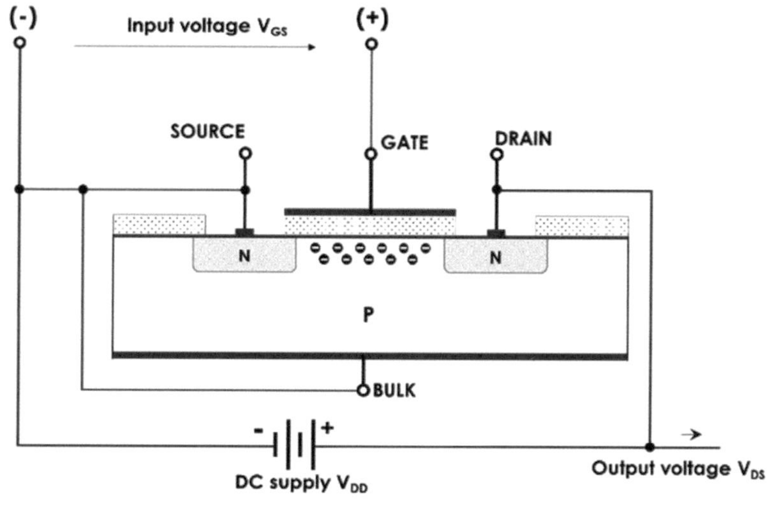


**Figure 2.**
*Operational supply of the metal-oxide-semiconductor field effect transistor (MOSFET) – n-channel.*

$R_{DS}$ or $R_{ON}$: the equivalent resistance at the output between drain and source pins that indicates the power output capability with a polarized voltage $V_{DS}$ chosen.

This basic structure of this MOSFET stays dedicated for medium or weak power capability because of the intrinsic values and limits of its dimensional and physical parameters. Also, in this topology, the path of the drain current stays "horizontal," parallel to the surface under the gate and in a width of some microns (μm).

## 2.2 Power MOSFET cell

To consider a real power MOSFET with high-level features of voltages and current supply, some technological improvements have been deployed to the initial generic structure of the MOSFET [6–8]. The new structure is now composed of four layers of alternating p-type and n-type doping. The path of conduction is now vertically polarized by the technology principle of a vertical diffused orientation for the channel (VDMOS) and the conditions of supply, as represented in **Figure 3**. The drain area is the based-bulk support of the global chip, at the bottom of the chip, and gate and source metallization are on the same plan on the surface (Top) of the chip.

Many elementary cells of this type are placed in parallel in the wafer, connected in the top plan by the source zones, so as to sum the currents at the global output of the device. A complete power MOSFET is composed of thousands of cells in parallel realized by this "source" connections. The realization of the shapes of these semiconductor zones has an impact on the total equivalent on-state resistance. In these structures, with many alternating doped layers, it appears also intrinsically undesirable construction of an n-p-n BJT. To avoid the case of this "parasitic" transistor that can be turned-on beyond conditions of polarization of the MOSFET, the source metallization overlaps onto the body doping region (channel) so as to shorten a part of the intrinsic BJT. This particularity is visible on the cross-section views of **Figures 3–5**. It stays only one junction, acting as a diode at the output drain – source, the cathode oriented at the drain and the anode on the source. This "parasitic" diode" can limit some applications, with its time of condition, and also be useful for protection in hard switching modes with a role of inverse rectifying.

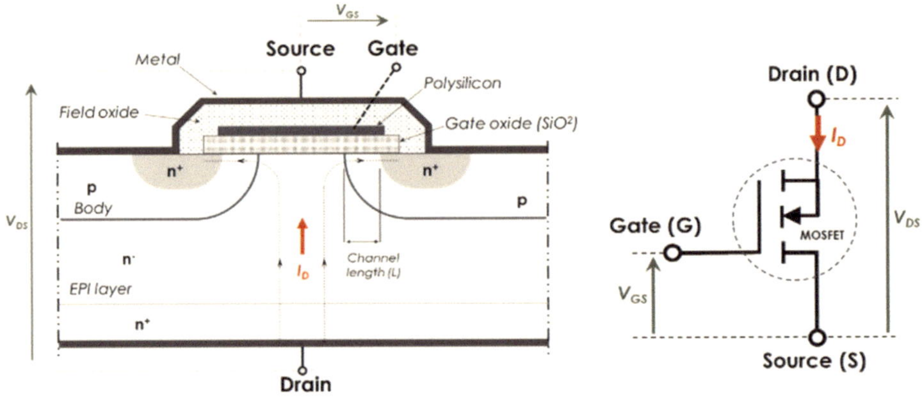


**Figure 3.**
*Cross section of the n-channel power MOSFET (vertical conduction of $I_{DS}$).*

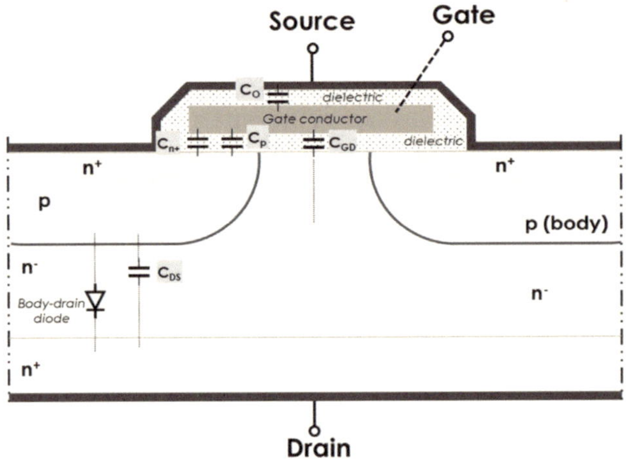


**Figure 4.**
*Origin of dynamic internal elements of the n-channel power MOSFET.*

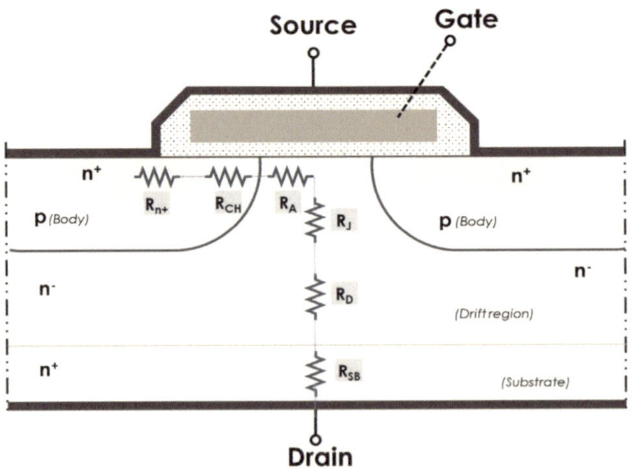


**Figure 5.**
*Origin of internal resistances of the n-channel power MOSFET in ON-STATE operation.*

We must also consider internal capacitive behaviors equivalent to the stockage or the transfer of charges between the gate, the drain, and the source layers and material disposition [9–11].

Two of these capacitors, $C_{GS}$ between gate and source and $C_{GD}$ between gate and drain, can be directly extracted from the geometry of the device.

Capacitance $C_{GS}$ is formed by the overlap of the source and the channel regions recovered by the gate electrode. Its definition and value depend mainly on the dimensions and geometry of the regions, and can be considered as the contributions of the three intrinsic capacitances $C_{n+}$, $C_O$, and $C_p$ all in a parallel disposition, as seen in **Figure 4**:

$$C_{GS} = C_{n+} + C_O + C_p \tag{1}$$

The expression of the capacitance $C_{GD}$ depends on two capacitive effects: the first one is due to the overlap of channel region and the gate electrode; the second one corresponds to the equivalent capacitance due to the depletion region varying with drain-source voltage (**Figure 4**). So, the general expression of $C_{GD}$ can be represented by the equivalent non-linear equation depending on the voltage $V_{DS}$:

$$C_{GD} = \frac{C_{GD,0}}{1 + K_1.\sqrt{V_{DS}}} \tag{2}$$

with $C_{GD,\,0}$ being theinitial value of $C_{GD}$ ($V_{DS} = 0$) and $K_1$ being the fitting parameter

The third is the one considered between the drain and source, $C_{DS}$. It corresponds to the defined junction capacitance of the parasitic body diode, inherent to the structure of vertical MOSFET (**Figure 4**). Its equation shows also a non-linear behavior among the voltage $V_{DS}$:

$$C_{DS} = \frac{C_{DS,0}}{K_2.\sqrt{V_{DS}}} \tag{3}$$

with $C_{DS,\,0}$ being the initial value of $C_{DS}$ ($V_{DS} = 0$) $K_2$ being the fitting parameter

In a MOSFET chip, equivalent resistive behaviors in the different metallic and semiconductors layers can be understood and synthetized with an ideal transversal cross-section view of the structure, represented at **Figure 5**. They are important to consider the performances of power losses and heating during switching loaded operations [12, 13]. At the semi-conductor level, we can consider the resistive contributions of the channel, of the accumulation layer, and of the drift region and the parasitic resistances of metallization, bond wires, and package.

When the transistor is conduction mode, the following contributions of resistive are as follows:

$$R_{DS(ON)} = R_{n+} + R_{CH} + R_A + R_J + R_D + R_{SB} \tag{4}$$

Rn+: diffusion source resistance of the n + region, very low contribution to the total value of $R_{DS(ON)}$.

$R_{CH}$: channel resistance, vary with geometry of the channel during operations.

$R_A$: accumulation resistance. When $V_{GS} > 0$, electric charges accumulate in the n-region near the gate area.

$R_J$: component resistance in the n + region between the two p-body regions

$R_D$: drift region resistance, localized in the main n-region, under the two p-body regions to the n + substrate region (drain). Main contribution to the global $R_{DS(ON)}$ at high $V_{DS}$.

$R_{SB}$: substrate resistance, important at lower voltages (<50 V).

Other contributions of parasitic elements can be added, if non-negligible in certain supply conditions, by the sum of bond-wire resistances, contact resistance (metallization, lead frame) and pins resistance and inductances of the package.

### 2.3 Main characteristics of MOSFET transistors

The current and voltage variations of the MOSFET transistor can be calculated and represented with generic analytical expressions issued from basic technological and semi-conductor properties [5, 7, 13]. For conventional operation in static and dynamic modes, some characteristics obtained at external connexions of the component in its package are more convenient for all final designers and users of MOSFET devices in power circuits and systems. The basic topology for this representation is to consider the transistor with the input at the gate pin, and the output at the drain one. The source pin will be considered as the common equipotential reference for the two other connexions, gate and drain. As input impedance of MOSFET is essentially capacitive, with a very weak leakage current on the gate, the main curve described and used in analytical and electrical model are the set of parametric representations of the drain current among the voltage command and supply: $I_{DS} = f(V_{DS}, V_{GS})$, and its derivatives: $I_{DS} = f(V_{GS})$ at $V_{DS}$ fixed, $R_{DS} = \delta V_{DS}/\delta I_{DS}$, and $g_m = \delta I_{DS}/\delta V_{GS}$.

### 2.4 Output characteristics I-V

The output characteristics of the MOSFET transistor, $I_{DS} = f(V_{DS}, V_{GS})$, are represented with variations of current $I_{DS}$ among output voltage $V_{DS}$, parametrized by different values of the input voltage $V_{GS}$ [14]. With this, all the different cases of operation are identified: the cutoff mode, the linear (ohmic) region, and the saturation (active) region, that are visible in **Figure 6** that represents the ideal set of $I_{DS}$ curves for MOSFET transistors. Two electrical parameters are defined either from physical, analytical, or empirical calculations or characterizations that delimits these three modes of operation [15]:

The threshold voltage, denoted as $V_{TH}$

The saturation voltage, denoted as $V_{DSSAT}$

So, the complete behavior of the transistor is described by the following:

For $V_{GS} < V_{TH}$: cutoff mode, then $I_{DS}(V_{GS}, V_{DS}) \approx I_{OFF} \approx 0$

For $V_{GS} > V_{TH}$: conduction mode, then $I_{DS}(t) = f[V_{GS}(t), V_{DS}(t)]$

for $V_{DS} < V_{DSSAT} = (V_{GS}\text{-}V_{TH})$, then $I_{DS}(V_{GS}, V_{DS}) \approx K^*V_{DS}$: linear or ohmic mode

for $V_{DS} > V_{DSSAT} = (V_{GS}\text{-}V_{TH})$, then $I_{DS}(V_{GS}, V_{DS}) \approx I_{DS}(V_{GS}) \approx I_{DSAT}(V_{GS})$: saturated mode, the current $I_{DS}$ is independent of the variation and the increase of $V_{DS}$. At high $V_{GS}$ ($\approx V_{DD}$ or $V_{DSMAX}$), the single value of $I_{DSAT}$ is called $I_{ON}$.

Some critical parameters are presented in these curves that help to understand and control the limits of operation modes of a MOSFET transistor [16]:

- The breakdown voltage, denoted $BV_{DSS}$, indicates the maximum drain-source voltage that can be applied before the mechanism of avalanche breakdown in the drain-body junction. In this case, the drain current arises dramatically, and an

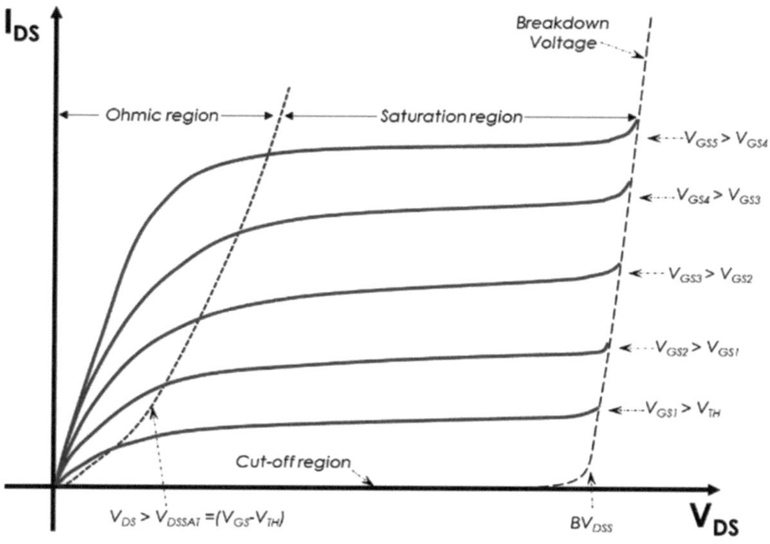


**Figure 6.**
*Output characteristics of a MOSFET transistor.*

instantaneous high-power dissipation leads to an uncontrolled high temperature step and the destruction of the chip.

- When $V_{DS} > V_{DSSAT}$ = ($V_{GS}$-$V_{TH}$), the current $I_{DS}$ is practically independent of the voltage $V_{DS}$. It is acting as an ideal current source, controlled by the voltage on the gate $V_{GS}$. This region is also called "active region" instead "saturation region," just to avoid a confusion with the same behavior in BJT. At exactly $V_{DS}$ = $V_{DSSAT}$, the boundary curve between ohmic and saturation region is given in a first-order analytical expression by the following:

$$I_{DS(ON)} = K.V_{DS}^2 \tag{5}$$

with K, a constant parameter issued from device geometry [17, 18].

## 2.5 Transfer characteristics

The conventional transfer characteristic for MOSFET transistors is the representation of the variation of current $I_{DS}$ at the output among the control voltage $V_{GS}$ at the gate. This representation introduces the concept of transverse gain or transconductance $g_m$. This parameter quantifies the change in drain current divided by the change in gate voltage for a constant drain voltage. So, it can be defined by the derivative of $I_{DS}$ variations among the voltage gate $V_{GS}$ at $V_{DS}$ = constant

$$g_m = \frac{\delta I_{DS}}{\delta V_{GS}}\rfloor_{V_{DS}=constant} \tag{6}$$

For conventional representation, $I_{DS}$ = f($V_{GS}$) is represented in the active or saturation region, so the fixed value of drain voltage must be chosen superior at the limit $V_{DSSAT}$ = ($V_{GS}$ − $V_{GSth}$). The slope of the transfer curve, that is, the transconductance $g_m$, presents two regions of definition:

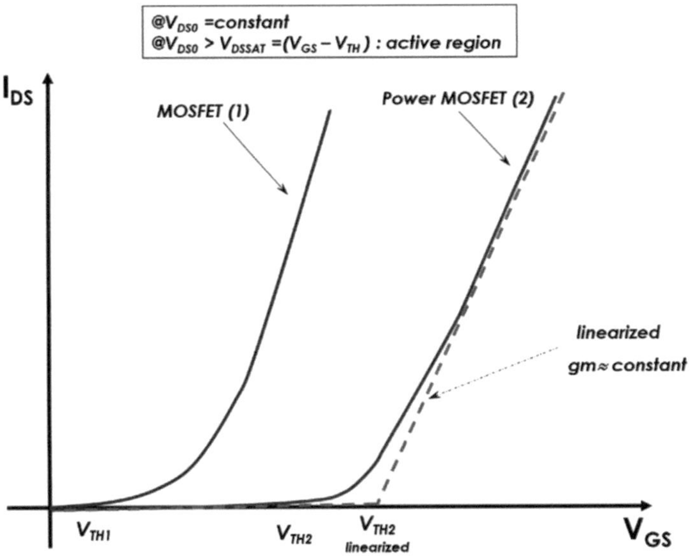


**Figure 7.**
*Typical transfer characteristics $I_{DS}$ $(V_{GS})$@$V_{DS}$ = constant of MOSFET transistors.*

for $V_{GS} < V_{TH}$, $I_{DS} \cong I_{OFF}$ and gm is very low or undetermined ($\cong 0/\infty$)

for $V_{GS} > V_{TH}$, $g_m = f(V_{GS})$ with a formal parabolic variation($\approx V_{GS}^2$). This is valid for low-signal MOSFET, where the normal operation voltage $V_{DS}$ stays under logical levels (5 V) or low-power levels (20 V).

For MOSFET in hard switching uses and voltage levels of operation >20 V, we observe, at a functional value of $V_{DS} > V_{DSSAT} = (V_{GS}\text{-}V_{GSth})$, a quasi-linear variation of the drain current with the gate voltage [19]. In this case, the transfer output characteristics ($I_{DS}$ = f $V_{GS}$) can be linearized, and the equivalent transconductance $g_m$ stays quite constant (**Figure** 7). This quantity is often defined by manufacturers at $V_{DSO}$, half of the breakdown voltage $B_{VDSS}$. Nevertheless, to obtain a high current handling capability with minimum gate drive voltage, large values of transconductance are a key issue for power MOSFET operability.

$$I_{DS} = g_{m.} V_{GS_{(V_{DS}=constant)}} \tag{7}$$

## 2.6 Drain resistance characteristics

Another critical and important parameter for normal operation of MOSFET with power features is the output equivalent resistance $R_{DS}$ and especially during conduction where it is denoted as $R_{DSON}$. The ON resistance determines the conduction power dissipation and impacts the losses in hard switching modes [20, 21]. It is defined by the different regions of the chip where flows the output current, as described previously. It is also an analytical calculation of the output characteristics $I_{DS} = f(V_{DS})$, where we can extract the slope by the following derivation:

$$R_{DS} = \frac{\delta V_{DS}}{\delta I_{DS}} \rfloor_{V_{GS}} \tag{8}$$

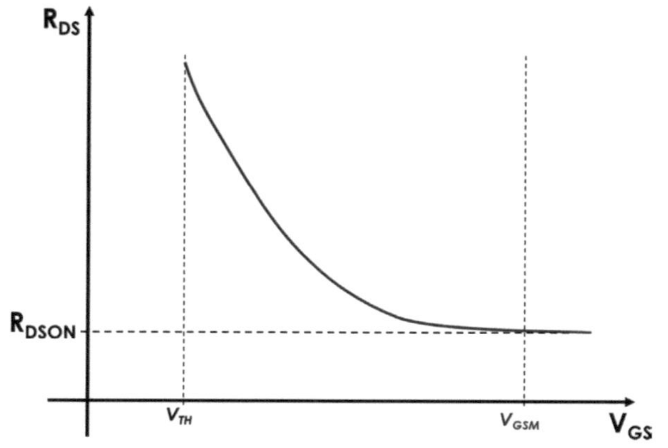


**Figure 8.**
*Global variation of the drain resistance $R_{DS}$ beyond the gate voltage $V_{GS}$.*

So, by technological concepts on the layers of semiconductor, it is normal that this resistance presents large variations among the conditions of supply and more especially with the shift of gate voltage $V_{GS}$ (**Figure 8**).

In cut-off region, when $V_{GS} < V_{TH}$, $R_{DS}$ trends to high values that considers the output as an open circuit.

In conduction mode, we can describe the variation of $R_{DSON}$ among two regimes:

In ohmic (linear) region, $I_{DS} = f(V_{DS})$ curves are linear, and slope for equivalent resistance $R_{DSON}$ is constant, but with a different value for each condition of gate voltage supply.

In active (saturation) region, current drain is practically constant among the arise of $V_{DS}$, so $R_{DSON}$ is minimum with values that can be under 1 ohm.

Another view to control the operation of MOSFET is to consider that the device is equivalent to an output resistance $R_{DS}$ controlled by the voltage at the input. There are also some physical considerations, among the voltage of the output, that explains the variations of $R_{DS}$ with the different bodies of the layers of the semiconductor structure of the MOSFET. For a gate voltage fixed, in normal conduction state ($V_{GS} > V_{TH}$), $R_{DS}$ vary with the current into the drain, as represented in **Figure 9**. In high-voltage MOSFETS, where the drift region resistance dominates the global $R_{DSON}$, the impact of the current drain on resistance shifts is weak. Finally, the most critical effect on $R_{DSON}$ is the strong dependency with the temperature [22]. The temperature coefficient of $R_{DSON}$ is always positive because of the majority-only carriers in the conduction mode of the MOSFET. For a rise of temperature of 100°C (25–125°C), the value of the on-resistance can double.

## 2.7 Transient and switching characteristics

### *2.7.1 Capacitive effects*

In a MOSFET, transient and switching modes are driven by the three equivalent capacitances at the terminals of the chip [3, 11]: $C_{\text{Gate-Source}}$ ($C_{GS}$), $C_{\text{Gate-Drain}}$ ($C_{DS}$), and $C_{\text{Drain-Source}}$ ($C_{DS}$).

The gate-source capacitance $C_{GS}$ is considered as weakly sensitive to electrical conditions of supply, as $V_{DS}$ variation.

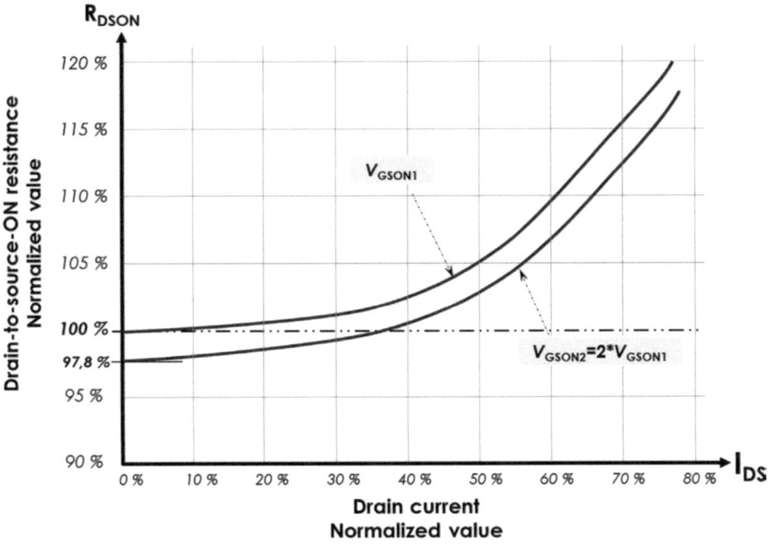


**Figure 9.**
*Representations of the variation of output resistance $R_{DSON}$ beyond the output current $I_{DS}$.*

The gate-drain capacitance $C_{GD}$ is very dependant of the supply voltage $V_{DS}$. At low voltage $V_{DS}$, its initial value is like $C_{GS}$. With the $V_{DS}$ voltage arise, $C_{GD}$ decreases to very weak value to reach the cut-off state

The drain-source capacitance $C_{DS}$ is basically a junction capacitance. Its value is inversely proportional to the voltage $V_{DS}$. Its role can be considered negligible in normal operation because of its effects hidden by the main impact of $C_{GD}$. Nevertheless, there are some specific switching modes where $C_{GD}$ is weak and $C_{DS}$ remains highlighted.

The $C_{GD}$ capacitance is the major term for determining rise and fall times of the chip with a circuit switching configuration. It is acting as a reaction between the input and the output, so its transient and frequency role can be highlighted at some operating conditions (Miller effect).

### *2.7.2 Dynamic and switching characteristics*

The dynamic behavior and performance of a MOSFET transistor are dependant of the intrinsic capacitances and resistances. For switching operations, with a two-state signal voltage at the input ($V_G$), output switching characteristics are established by the gate charge description and the reverse recovery role of the body diode [3, 5]. The different time-steps for a basic ON-OFF state of a MOSFET are denoted:

$t_{DON}$: delay for $V_{GS}$ rise from $V_{LOW}$ (0 V) to $V_{TH}$. No drain current $I_{DS}$.

$t_R$: delay for $V_{GS}$ rise from $V_{TH}$ to $V_{GS\ ON}$ (full charged gate). Arise of drain current $I_{DS}$.

$t_{ON}$: delay for $V_{GS}$ rise from $V_{GS\ ON}$ to $V_{HIGH}$. Constant drain current $I_{DS} = I_{DS\ ON}$

$t_{DOFF}$: delay for $V_{GS}$ decrease from $V_{HIGH}$ to $V_{GS\ ON}$. Decrease of drain current $I_{DS}$.

$t_F$: delay for $V_{GS}$ decrease from $V_{GS\ ON}$ to $V_{TH}$. Decrease of drain current $I_{DS}$ to null.

$t_{VOFF}$: delay for $V_{GS}$ decrease from $V_{TH}$ to $V_{LOW}$ (0 V). No more drain current $I_{DS}$.

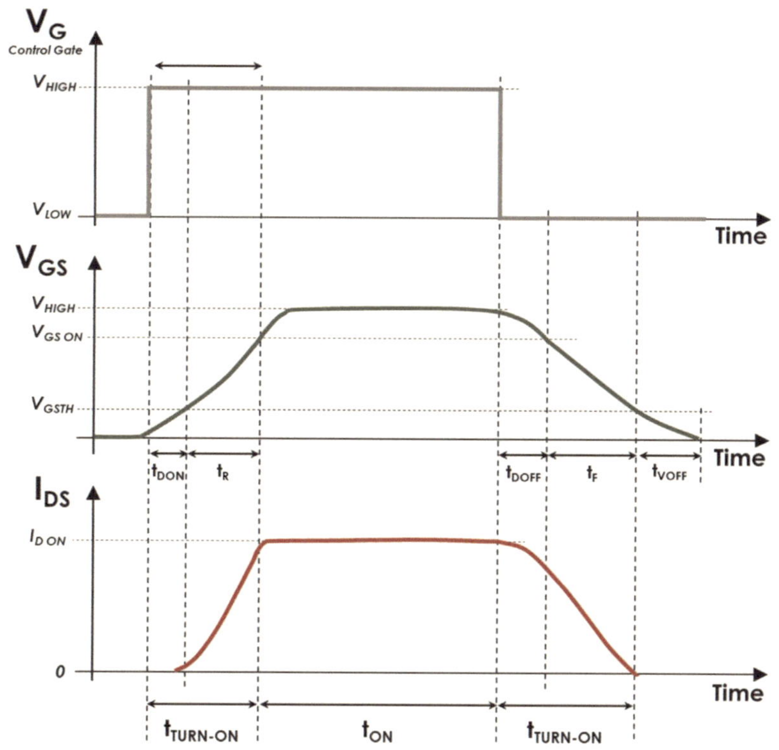


**Figure 10.**
*Basic gate charge waveforms of Power MOSFET during turn-on and turn-off transitions.*

**Figure 10** represents the different waveforms in time-domain and the related time steps for an ideal sequence of ON-OFF switch command at the external gate terminal of the transistor.

### 2.8 Other main critical parameters

#### *2.8.1 $V_{TH}$: Gate threshold voltage*

This is the intrinsic gate voltage at the input that initiates drain current flow at the output. An ideal value for MOSFET operations is a minimum one, to reduce turn-on delay time [4, 6, 7]. But for Power MOSFET, there are some compromises between a lower value of $V_{TH}$ and failures cases that can occur during high levels switching modes, because of the thin oxides and voltage rating. For all MOSFET devices, the inner threshold voltage spreading is specified by minimum and maximum values. Common values are 1–2 V for medium voltage devices (200 V) and 2–5 V for high voltage devices (1200 V).

#### *2.8.2 $V(BR)_{DSS}$: Drain-source breakdown voltage*

This is the drain-source voltage supported by the doped Epi layer before entering into avalanche mode and uncontrolled drain current rising. It is denoted $V(BR)_{DSS}$ (or $BV_{DSS}$) and is defined for a specified leakage current $I_{DSS}$ and a fixed temperature because of its strong positive coefficient temperature.

#### *2.8.3 $I_{DSS}$: Zero gate voltage drain current*

This is the drain-source leakage current with a gate-source voltage $V_{GS}$ = 0 V. Its value is specified for a fixed Drain-source voltage $V_{DS}$ and a range of temperature. Leakage power loss is $I_{DS}$S times drain-source voltage and is usually negligible.

#### *2.8.4 $I_{GSS}$: Gate-source leakage current*

This is the leakage current that flows through the gate terminal at a specified gate-source voltage.

## 3. Review of MOSFET transistor modeling

### 3.1 Basic model of MOSFET

Basically, at the semiconductor or chip level, electrical models of MOSFET transistors are issued from analytical and physical equations at the semiconductor level [2, 9, 13, 21]. The equivalent electrical model of generic MOSFET can be considered by representing the amount of drain current flow controlled by voltage sources or potentials ($V_{GS}$ and $V_{DS}$), and a set of capacitive elements to reproduce the transient characteristics and the time delays during time-varying and switching operations. A model of a junction can also be included between drain-source terminals, to consider the non-negligible effects of leakage currents with the reverse body diode, especially in power MOSFET semiconductors. These basic components and their placement in the electrical schematic of the global model are represented in **Figure 11**.

With many years of works, adaptations and improvements on the topology and on the modeling way of MOSFET transistors, a set of generic MOSFET models have been well established and normalized for implementation in electronic solvers (SPICE) [23, 24]. For users that have no access at internal physical parameters of the semiconductors and the materials, funders and manufacturers propose their standardized SPICE models dedicated to their chips, with a good compromise between analytical/empirical technical information enclosed for efficient simulations [25, 26].

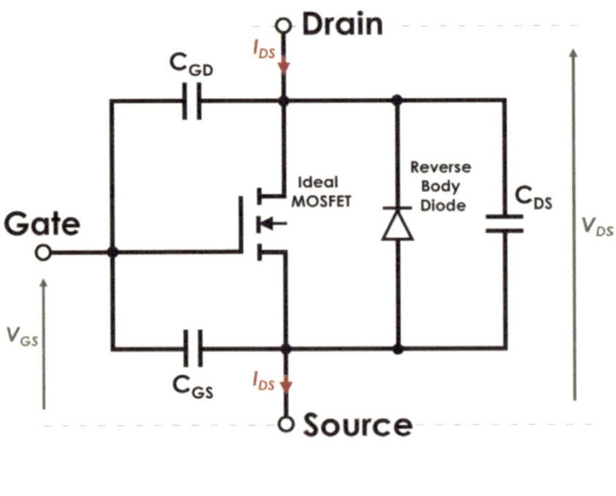


**Figure 11.**
*Basic generic electrical model for MOSFET structure.*

### 3.2 Model of the equivalent current source $I_{DS} = f(V_{GS}, V_{DS})$

The "ideal" MOSFET model represents the variations of the drain current in the three regions of operation: cutoff, linear or ohmic, and saturation or pinch-off. Based on some technological and physical parameters of the chip, these expressions are typically implemented in electrical circuit solver (SPICE), by the set of following basis equations, with $V_{TH}$ the threshold voltage at the gate-source junction:

*Cutoff mode:* $V_{GS} < V_{TH}$

$$I_{DS}(V_{GS}, V_{DS}) = 0 \tag{9}$$

*Conduction mode:* $V_{GS} > V_{TH}$

If $V_{DS} < (V_{GS} - V_{TH})$, then

$$I_{DS}(V_{GS}, V_{DS}) = K * \left[ (V_{GS} - V_{TH}) * V_{DS} - \frac{V_{DS}^2}{2} \right] \tag{10}$$

The device is denoted in the linear mode

If $V_D > (V_{GS} - V_{TH})$, then

$$I_{DS}(V_{GS}, V_{DS}) = \mathrm{K} * \left[ \frac{(V_{GS} - V_{TH})2}{2} \right] \tag{11}$$

The device is denoted in the saturated mode: $I_{DS}$ is practically independent of the $V_{DS}$ variation.

The parameter K used in $I_{DS}$ equations is called the intrinsic transconductance (A/$V^2$). It is obtained from technological considerations of the semiconductor device and expressed by

$$K = \mu_n . C_{OX} . \frac{W}{L} \tag{12}$$

with

$\mu_n$ being the mobility carrier (electrons) in $m^2$/ (V.s)

$C_{OX}$ being the gate oxide capacitance in F/$m^2$

W being the width of the channel in m

L being the length of the channel in m

Cox is the equivalent oxide capacitance calculated by the normalized square formula:

$$C_{OX} = \frac{\varepsilon_{r\ SiO^2} . \varepsilon_0}{t_{OX}} \tag{13}$$

With

$\varepsilon_{r\ SiO}{}^2$ being the dielectric constant of $SiO^2$ oxide (3.9)

$\varepsilon_0$ being the absolute permittivity of air ($\varepsilon_0$ = 8.85 pF/m)

$t_{OX}$ being the thickness of the oxide $SiO^2$

This model of the drain current $I_{DS}$ is a first-order approximation of long-channel device (L > 10 μm). Equations are based on the Shichman-Hodges model [2] and referenced as LEVEL 1 model in solvers. An additional parameter, λ (Lambda), is used for the channel length modulation effect that represents the rise of drain current with

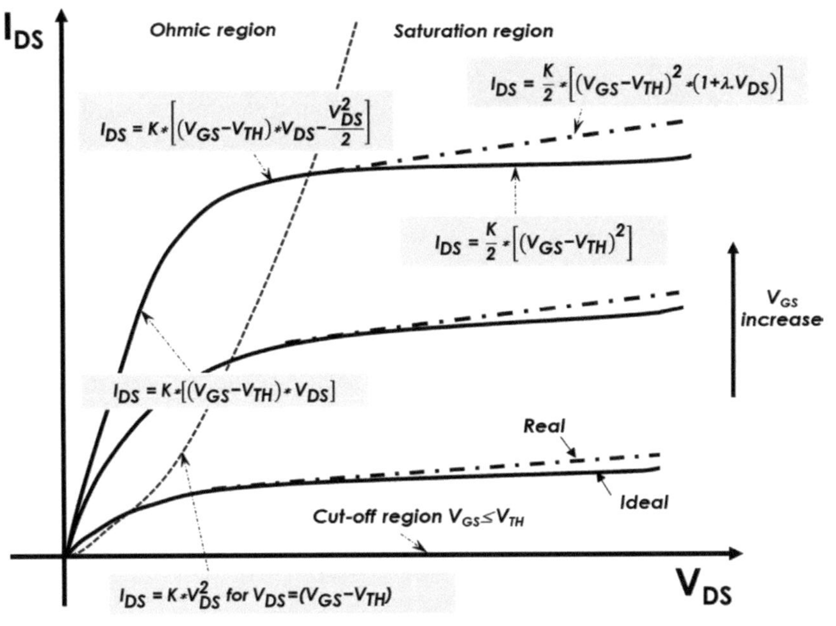


**Figure 12.**
*Overview of basic analytical/electrical modeling of the output drain current in MOSFET devices.*

$V_{DS}$ in the saturated mode and shows a (slow) positive slope in output characteristics instead of a constant current $I_{DSSAT}$:

Ohmic mode:

$$I_{DS}(V_{GS}, V_{DS}) = K * \left[(V_{GS} - V_{TH}) * V_{DS} - \frac{V_{DS}^2}{2}\right] * (1 + \lambda.V_{DS}) \tag{14}$$

Saturated mode

$$I_{DS}(V_{GS}, V_{DS}) = K * \left[\frac{(V_{GS} - V_{TH})2}{2}\right] * (1 + \lambda.\mathrm{V_{DS}}) \tag{15}$$

The following figure (**Figure 12**) represents the synthetises of the calculations of the $I_{DS}$($V_{GS}$ and $V_{DS}$) curves, with qualitative details on the sensitivity and impact of parameters of the equations on the curve shapes and the different zones of the MOSFET operations.

This initial set of analytical equation of $I_{DS}$ in the conduction mode has been refined in successive modeling steps based on semi-empirical and experimental observations during the evolutions and the development of MOSFET transistors. Improvements are consequent on trending to develop more behavioral and macro-models, when physical and technological input parameters are often not directly and simply available and because of the increasing complexity of the chips [27, 28]. To synthetize, different levels of the model of MOSFETs have been implemented and are available in SPICE solvers:

- The first-generation models, from Level 1 to Level 3, and upper that include technology progress (reduction of the dimensions) and empirical effects, as junction temperature and threshold voltage shifts.

- The second-generation models for compliance with deep-submicron MOS technology. This semi-empirical modeling process has been initiated and is continually updated by the Berkeley's BSIM group, from University of California. Actual version (BSIM4.xx) is commonly used in advanced VLSI design and industry.

- The third-generation models, derived from the first ones but adapted with behavioral "external" representation of new capabilities and features of MOSFET transistor, especially for high-voltage, high current and high-power operations. This generation trends to achieve the robust physical-based macromodel that could simulate preventive cases of failures and out-of-range operations of the device with multiple stresses and hard switching conditions [29].

The "final" generation for a versatile, smart, and efficient model of MOSFET chips with their package and their routing connections would be the "Digital Twin" model [30, 31].

### 3.3 Modeling the capacitances

Constant value or variable expressions of the three capacitance $C_{GS}$, $C_{GD}$, and $C_{DS}$ complete the initial model. Rigorously, they are not constant but can vary with internal electric fields induced by voltage variations at the gate, drain and sources pins. Usually, it is considered that capacitance $C_{GD}$ has the main significant change with $V_{DG} \approx V_{DS}$, presenting a large non-linear variation about a factor of 10 to 100. $C_{GS}$ can be considered constant, and $C_{DS}$ is negligible for its effect on time and delay in normal operation of MOSFET transistor. Typical variations of MOSFET capacitances with voltages are represented at **Figure 13**.

Funders and manufacturers can provide an access to initial global values of the intrinsic capacitances by calculations or measurements of external capacitances at the ports of the chip, denoted $C_{ISS}$, $C_{OSS}$, and $C_{RSS}$.

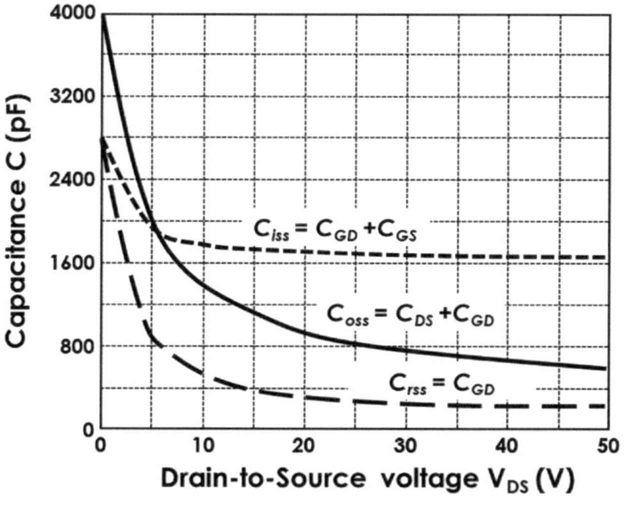


**Figure 13.**
*Typical variations of MOSFET capacitances with drain voltage $V_{DS}$.*

- The input capacitance, $C_{iss}$, is measured between the gate and source terminals, with a shortened condition at the output $V_{DS\_AC}$ = 0 V. So, $C_{iss}$ is the global value of gate-drain capacitance $C_{GD}$ and gate-to-source capacitance $C_{GS}$ placed in parallel.

$$C_{Gate-Source}|_{V_{\tilde{DS}=0V}} = Ciss(V_{DS\_DC}) = C_{GS}(V_{DS\_DC}) + C_{GD}(V_{DS\_DC}) \tag{16}$$

This input capacitance has a main impact on the turn on and turn off delays. To a normal switching activity, it must be charged to the threshold voltage for the start of the turn-on of the device and then discharged to the low-state control voltage to turn off entirely the transistor before the next switching action.

- The measurement of output capacitance between drain and source, with input gate-source at 0 V in AC, gives the capacitance denoted $C_{oss}$. It is the equivalent capacitance of $C_{GD}$ and $C_{DS}$ in parallel.

$$C_{Drain-Source}|_{V_{\tilde{GS}=0V}} = Coss(V_{DS\_DC}) = C_{DS}(V_{DS\_DC}) + C_{GD}(V_{DS\_DC}) \tag{17}$$

$C_{OSS}$ has an important impact on soft switching command because of its role on resonance of the circuit with loads.

- The measurement of reverse or transfer capacitance between drain and gate, source at 0 V in AC, gives the capacitance denoted as $C_{rss}$. In this configuration, the $C_{RSS}$ capacitance is the direct measure of the gate-to drain capacitance $C_{GD}$.

$$C_{Gate-Drain}|_{V_{\tilde{S}=0V}} = Crss(V_{DS\_DC}) = C_{GD}(V_{DS\_DC}) \tag{18}$$

### 3.4 A global model for power MOSFET cell

Many improved and dedicated models of MOSFET and power MOSFET have been developed, starting on the generic circuit and equations presented before. For power MOSFET, there is some particular attention with the equation of drain current $I_{DS}$, because of intrinsic non-linear variations of the transconductance in high level switching modes, and inherent impacts of internal temperature rises on electricals output characteristics of the device. In the following, we propose a compact and robust analytical equation of $I_{DS}$ for resolving these critical limits of MOSFET models.

### 3.5 Topology of a generic model for power MOSFET

The main region of operation of Power MOSFET is the on–off states, with drain-source voltage $V_{DS}$ polarized in the saturated region ($V_{DS} >> V_{DSAT}$), and the command $V_{GS}$ having only two states: $V_{GS} < 0$ and $V_{GS} >> V_{th}$. The topology proposed is represented at **Figure 14**. This electrical schematic is composed by a current source $I_{DS}$ controlled by gate voltage $V_{GS}$ with the transconductance $g_m$ (A/V), a resistor $R_{DSON}$ for losses representation at on-state transistor, by the voltage source $V_{GSth}$ that represents the threshold voltage $V_{TH}$ activating the different states of the chip. The schematic is completed by three global capacitors, $C_{GS}$, $C_{DS}$, and $C_{GD}$, for the parasitic and time charge effects, and by diodes $D_{DS}$ and $D_G$ for the "body" diode protections against return. All the parameters into these components could be determined either by semiconductor physical parameters, if available, either by dedicated external characterization, given by the datasheet of the manufacturers, and/or by own user

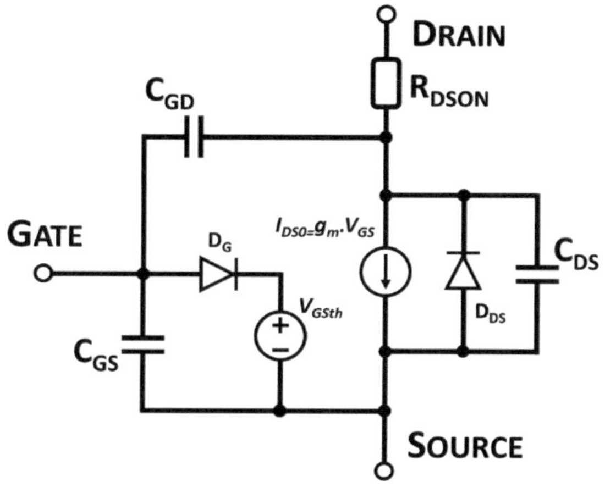


**Figure 14.**
*Topology of the proposed circuit model for power MOSFET chip.*

characterization. For this, protocols and conditions of the characterizations for parameter extraction are critical for initial realistic values and for correct identification with the final parameters and equations of the model.

This equivalent model of MOSFET chip uses eight equivalent electrical components listed in **Table 1**: two sources, two junctions, one resistor, and three capacitors. The parameters and their value can be constant or variable, among the voltages, currents, frequencies conditions of supply and excitations. Most semiconductor's equivalent components can also be highly dependent on internal and external temperature. This behavior is important and reinforced for power applications, where internal and external current variations can be amplified during normal and boundary cases of power supply. We propose to consider firstly ambient temperature to define and implement these components in the model, as described in the following section.

### 3.6 Non-linear model of transfer characteristic $I_{DS}(V_{GS})$

Fine modeling and representation of the transfer characteristic $I_{DS}(V_{GS})$ is the key for trusty calculation and simulation of MOSFET features because of its strong and

| Component label | Name | Role/effect |
|---|---|---|
| $g_m$ | Transconductance (A/V) | *Current drain control* |
| $R_{DSON}$ | Drain resistance (Ω) | *Conductive losses at ON-state* |
| $C_{GD}$ | Gate-drain capacitance (F) | *Gate charge delay* |
| $C_{GS}$ | Gate-source capacitance (F) | *Input switching delay* |
| $C_{DS}$ | Drain-source capacitance (F) | *Output charge delay – low effect* |
| $V_{GSth}$ | Threshold voltage source (V) | *Input voltage fall at ON-state* |
| $D_{DS}$ | Body drain diode | *Output junction – voltage rating* |
| $D_{DS}$ | Input junction diode | *Current gate $I_{GG}$ control at ON-state* |

**Table 1.**
*The eight components of a global model for power MOSFET chip in off-on operations.*

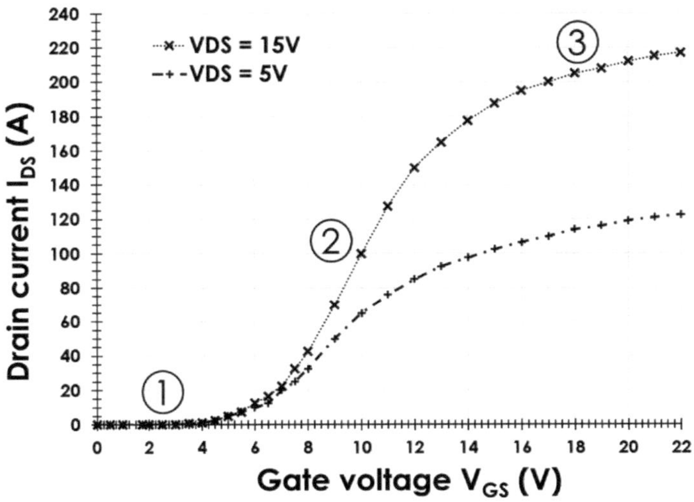


**Figure 15.**
*Experimental variation of drain current as a function of gate voltage in a power MOSFET device [31, 32] with the three zones: 1 – off zone, 2 – linear zone, and 3 – flat zone.*

complex non-linearity dependency to main physical parameter of the semiconductor: Vth, $R_{DS}$, carrier mobility ($\mu_{eff}$), channel effective length $L_{eff}$, etc. ...

Another representation is the variation (derivative) of the drain current $I_D$ among the gate voltage $V_{GS,}$ at a nominal fixed voltage $V_{DS}$, that defines the transconductance $g_m$ (A/V). These curves can be usually obtained experimentally either by the founder or the user, at specific conditions of supply and operations. The transfer characteristics $I_{DS} = f(V_{GS})@V_{DS0}$ presents generally same non-linearity variations that can be analyzed with three distinct zones (**Figure 15**).

First zone, for VGS < VTH, is the low-current source activity, with a flat low-level variation of $I_D$ among $V_{GS}$. Equivalent $g_m$ (the slope) starts from very low value but increases rapidly with $V_{GS}$.

The second zone, where $V_{GS}$ > VTH, shows a rapid quasi-linear arise of $I_{DS}$ with $V_{GS}$. In this zone, we have the current controlled source effect of the transistor. The transconductance $g_m$ has reached its maximum value and stays quite constant.

The last (third) zone is the flat one, starting about $V_{GS}$ > > VGSth, where $I_D$ becomes quite constant, at a maximum current denoted $I_{DSS}$, even if $V_{GS}$ still increases. Transconductance $g_m$ in this third region decreases from its maximum to lower values.

The $I_{DS}(V_{GS})$ curve is also dependant of the potential $V_{DS}$ applied at the output that affects mainly the maximum current $I_{DSS}$ and presents shifts both in position and in the slope ($g_m$) with different $V_{DS0}$ (**Figure 15**). It appears that the equivalent transconductance $g_m$ vary non-linearly with two electrical conditions: the applied voltage $V_{GS}$ at the gate of the transistor and the drain source voltage $V_{DS0}$.

Many works have proposed analytical expressions to reproduce these family of curves, and it has been discussed synthetically in previous paragraph, with different progress and levels in MOSFET model. The different technics and models proposed for this must comply and correspond to the real final use of the device. In this review on MOSFET model, we have been concerned by switching operations of high-power transistor, to drive the electrical motor, for example [33]. The objective is to operate at a $V_{DS}$ voltage that places the transistor in saturation region ($V_{DS0} >> V_{DSSAT}$), and the command $V_{GS}$ drives two "extreme" states of the device:

$$V_{GS} << V_{TH} \rightarrow \text{OFF state} \rightarrow I_{DS} = 0, V_{DS} = V_{DD} \text{ or } V_{SUPPLY}.$$

$$V_{GS} >> V_{TH} \rightarrow \text{ON state} \rightarrow I_{DS} = I_{DSSAT}\ (V_{GS} - V_{TH}) = I_{ON\ MAX}, V_{DS} \approx 0\ \text{V}.$$

The need for $I_{DS}(V_{GS})$ modeling is to reproduce the more accurately possible the drain current variations from zone 1 ($I_{OFF}$) to zone 3 ($I_{ON\ MAX}$). One of the more representative functions for the shape of the transfer characteristic is the mathematical function called "sigmoid," represented at **Figure 16**. This sigmoidal function is analytically denoted σ(x), with x an argument in the real domain, and its waveform is very similar to the three non-linear zones of transfer curves in MOSFET. The normalized expression of $\sigma(x) = \frac{1}{1+e^{-x}}$ is compared to a representation of the $I_{DS}(V_{GS})$ curve, where $\sigma \cong I_{DS}$ and $z \cong V_{GS}$. Normalized sigmoid amplitude is confined between 0 at x negative, and trends asymptotically to 1 when $x >> 0$. The slope of the curve is exactly centred to 0.5 when x = 0. We retrieve in this ideal function the three regions of typical variation curves of MOSFET transistors (**Figure 16**).

As the drain current equation among the gate voltage is representative of this variation, we can formally identify the transconductance curve by this sigmoidal function, with additional coefficients to have a set of functions parametrized by the actual electrical conditions of the drain current equation. The sigmoidal function can be generalized with extra coefficients denoted k, A, and r:

$$\sigma(x) = \frac{k}{1 + A\, e^{-r.x}} \tag{19}$$

Numerical values of k, A and r of this generalized sigmoidal function allow to adjust sigmoidal shape to similar curves with the following effects illustrated at **Figure 17**:

- k is acting on the amplitude gap of the sigmoidal function in vertical axis (**Figure 17a**)

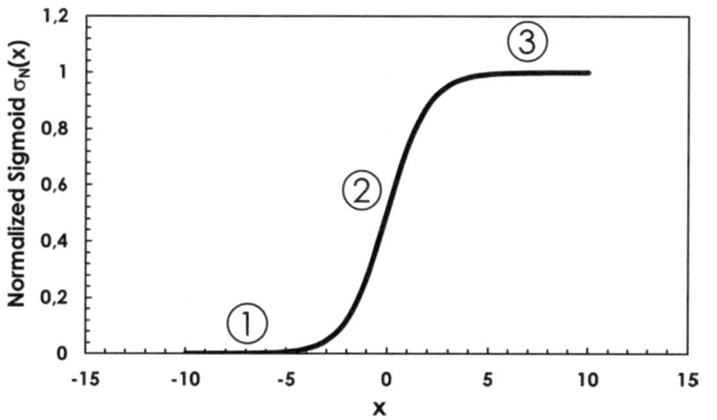


**Figure 16.**
*Variation of normalized sigmoidal function* $\sigma(x) = \frac{1}{1+e^{-x}}$.

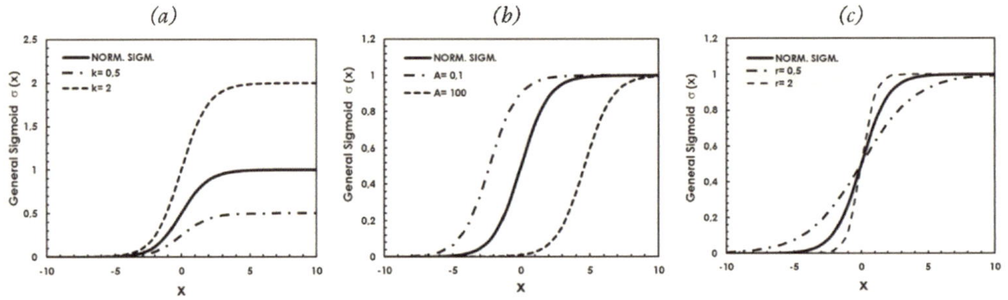


**Figure 17.**
*Variations of generalized sigmoidal function* $\sigma(x) = \frac{k}{(1+Ae^{-rx})}$ *(a) k variation; (b) A variation; (c) r variation.*

- coefficient A allows to shift the sigmoid curve among the x-axis (**Figure 17b**)
- r modifies the slope of the sigmoid, with the inflection point centred in x = 0 (**Figure 17c**)

We adapt the non-linear equation of drain current among gate voltage to this general sigmoidal equation, with the corresponding terms:
$x \cong V_{GS}$; $\sigma \cong I_D$; k, A, and r $\cong$ set of values for each $V_{DS}$ nominal fixed value ($V_{DS0}$).

$$I_{DS}(V_{GS})\rfloor_{V_{DS0}} = \frac{k\rfloor_{V_{DS0}}}{1 + A\rfloor_{V_{DS0}}\, e^{-r\rfloor_{V_{DS0}} \cdot V_{GS}}} \tag{20}$$

With this expression, it is possible to adjust and fit experimental curves of drain current versus gate voltage parametrized by the drain supply voltage $V_{DS}$ and to reproduce the non-linear variations of the transfer characteristics and of the

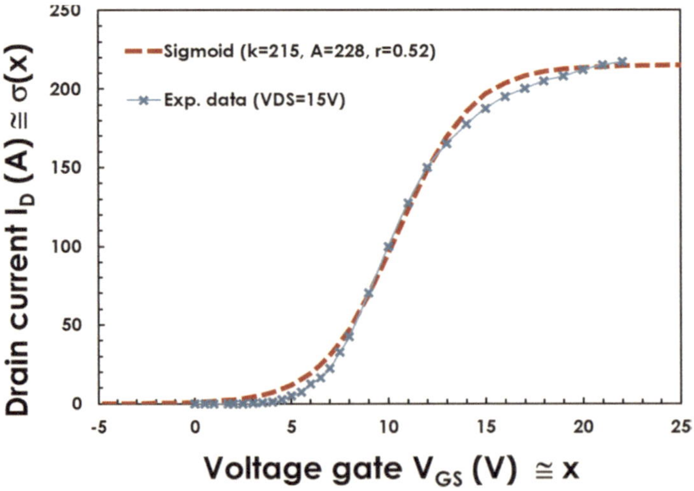


**Figure 18.**
*Comparison of transfer characteristic* $I_D = f(VGS)$ *measured and calculated with sigmoid function at* $V_{DS}$ = 15 V *for a power MOSFET transistor.*

transconductance of MOSFET. An example of this calculation is illustrated at **Figure 18**, where the experimental curve of $I_{DS}(V_{GS})$ of a Power Sic MOSFET is correctly fitted on all the nominal range of $V_{GS}$ voltage control.

The set of parameters (k, A, and r) are defined for a single voltage $V_{DS0}$ at the first order of identification, with the value of $V_{DS0}$ mainly in the saturation region of output characteristics $I_{DS}(V_{DS})$ with $V_{DS} > > V_{DSAT} = (V_{GS}\text{-}V_{TH})$. To expand the range of the model for simulation in different area of the transistor operation, many procedures of identifications of the parameters k, A, and r can be conducted at various values of the drain voltage. Then, a second-order modeling task is to define parametric variation of these three parameters among the $V_{DS}$ voltage applied and implement them in the transfer characteristic Equation [31]:

$$I_{DS}(V_{GS})\rfloor_{V_{DS}} = \frac{k(V_{DS})}{1 + A(V_{DS}).e^{r(V_{DS}).V_{GS}}} \tag{21}$$

This final equation proposed and applied to a MOSFET transistor has been proved to reproduce finely wide range of nonlinear behaviors of drain current among the voltages at the external terminal [33]. Only three empirical parameters, k, A, and r, are sufficient to reproduce MOSFET operations in normal operations and at extreme zones of switching cases. Once the main efforts for correct identification of the three model parameters have been performed, based on electrical characteristics and datasheet of the device, there are capacities to extend the set of parametric quantities as the temperature, so to be easily implemented in the $I_{DS}$ equation and finally to derive more robust and complete electrothermal model of power MOSFET [22, 31, 33].

## Author details

Jean-Marc Dienot[1,2]

1 University of Toulouse, France

2 Research Laboratory, SIAME-Labceem, France

*Address all correspondence to: jean-marc.dienot@univ-tlse3.fr

## References


[1] Shockley W. A unipolar field effect transistor. Proceedings of the IRE. 1952; **40**:1365-1376

[2] Schichman H, Hodges D. Modelling and simulation of insulated-gate field effect transistor switching circuits. IEEE Journal of Solid-State Circuits. 1968;**3**: 285-289

[3] Yang P, Epler BD, Chatterjee PK. An investigation of the charge conservation problem for MOSFET circuit simulation. IEEE Journal of Solid-State Circuits. 1983;**SC-18**(1):128-138

[4] SiC Power Devices and Modules: Application Note. ROHM Semicon. Co., Ltd. No. 63AN102E-Rev.003; 2020. Available from: fscdn.rohm.com/en/products/databook/applinote/discrete/sic/common/sic_appli-e.pdf

[5] Sze SM. Modern Semiconductor Device Physics. Wiley-Interscience; 1997. Available from: onlinelibrary.wiley.com. ISBN 0-471-15237-4

[6] Murari B, Bertotti F, Vignola GA. Smart Power ICs. Springer; 2002. Available from: https://www.springerpub.com/. ISBN 3-540-43238-8

[7] Jayant Baliga B. Fundamentals of Power Semiconductor Devices. Springer International Publishing; 2019. Available from: https://www.springerpub.com/. ISBN 978-3-319-93988-9

[8] Mohan N, Undeland TM, Robbins WP. Power Electronics. 3rd ed. John Wiley & Sons; 2003. Available from: https://www.wiley.com/en-fr/. ISBN 978-0-471-22693-2

[9] Wei L, Boeuf F, Skotnicki T, Wong H-SP. Parasitic capacitances: Analytical models and impact on circuit-level performance. IEEE Transactions on Electron Devices. 2011;**58**(5):1361-1370

[10] Lacord J, Ghibaudo G, Boeuf F. Comprehensive and accurate parasitic capacitance models for two- and three-dimensional CMOS device structures. IEEE Transaction on Electron Devices. May 2012;**59**(5):1332-1344

[11] Scott R, Franz A. An accurate model for power DMOSFETs including interelectrode capacitances. IEEE Transactions on Power Electronics. 1991; **6**(2):192-198

[12] Barkhordarian V. Power MOSFET Basics. El Segundo, USA. App Not. AN-1084: International Rectifier; 1997

[13] Leventhal R, Green L. Semiconductor Modeling. Springer Sciences; 2006. Available from: https://www.springerpub.com/. ISBN 0-387-24159-0

[14] Tsividis YP. Operation and Modeling of the MOS Transistor. MacGraw-Hill Edition; 1987. Available from: mheducation.com

[15] Jain S. Generalized transconductance and transresistance methods for MOSFET characterization. Solid-State Electronics. 1989;**32**(1):77-86

[16] Schoiswohl J. Linear Mode Operation and Safe Operating Diagram of Power-MOSFET. Munich, Germany. App. Not. AP990007-v1.1: Infineon Techn. AG; 2017

[17] Harper C. Electronic Material and Processes Handbook. McGraw-Hill; 2003. Available from: mheducation.com. ISBN 0-07-140214

[18] Neelakanta PS. Handbook of Electromagnetic Material. CRC Press;

1995. Available from: https://www.routledge.com/about-us/crc-press. ISBN 0-8493-2500-5

[19] Li H, Munk-Nielsen S. Detail study of SiC MOSFET switching characteristics. In: Proc. IEEE 5th Intern. Symp. PEDG. Galway, Ireland: IEEE (Proceedings); 2014. pp. 1-5. DOI: 10.1109/PEDG.2014.6878691

[20] Liu Y. Power Electronic Packaging. Springer Sciences; 2012. DOI: 10.1007/978-1-4614-1053-9. Available from: https://www.springerpub.com/

[21] Swaminathan M, Ege Engin A. Power Integrity Modeling and Design for Semi-Conductors and Systems. Prentice Hall; 2007. Available from: https://openlibrary.org/publishers/Prentice_Hall

[22] Cui Y, Chinthavali M, Tolbert LM. Temperature dependent Pspice model of silicon carbide power MOSFET. In: Proc. of 27th IEEE Appl. Power Electr. Conf. (APEC). Orlando, USA: IEEE; 2012. pp. 1698-1704. DOI: 10.1109/APEC.2012.6166050

[23] Fu R, Santi E, Zhang Y. Power SiC MOSFET model with simplified description of linear and saturation operating regions. In: Proc. 9th Intern. Conf. ICPE-ECCE Asia. Seoul, Korea (South): IEEE (Proceedings); 2015. pp. 190-195. DOI: 10.1109/ICPE.2015.7167785

[24] Johannesson D, Nawaz M. Analytical PSpice model for SiC MOSFET based high power modules. Microelectronics Journal. 2016;**53**:167-176. DOI: 10.1016/j.mejo.2016.05.001

[25] McNutt R. Compact Modeling of power devices. In: Extreme Environment Electronics Book. CRC Press; 2013. Available from: https://www.routledge.com/about-us/crc-press

[26] McNutt T, Hefner A, Mantooth A, Berning D, Singh R. Compact models for silicon carbide power devices. Solid State Electronics. 2004;**48**:1757-1762

[27] Santi E, Peng K, Mantooth HA, Hudgins JL. Modeling of wide-bandgap power semiconductor devices—Part II. IEEE Transactions on Electron Devices. 2015;**62**(2):434-442

[28] Huang S, Li X, Yeh T, Mao S, Sekisue T, Ambalavanar V. Multi-scale and multi-domain simulation of electrical power system. In: Proc. of IEEE Electron Dev. Techn. And Manuf. (EDTM 2017). Toyama, Japan: IEEE; 2017. pp. 100-102

[29] Khvitia B, Gheonjian A, Kutchadze Z, Jobava R. A SPICE model for IGBTs and power MOSFETs focusing on EMI/EMC in high-voltage systems. Electronics. 2021;**10**(2822):1-17. DOI: 10.3390/electronics10222822

[30] Batista E et al. Alstom multi-physics modelling methodology for railway electrical system. In: UGM ANSYS Conference. ANSYS S.A; 2011. Available from: https://www.ansys.com/events

[31] Dumollard Y, Batista E, Dienot JM, Pecastaing L. Equation-based modeling of the electrothermal behavior of a SiC MOSFET chip during a short circuit. In: 2020 IEEE Latin America Electron Devices Conference (LAEDC). San Jose, Costa Rica: IEEE; 2020. pp. 1-4. DOI: 10.1109/LAEDC49063.2020.9073318

[32] Semiconductor Datasheet. C2M0025120D - Silicon Carbide Power MOSFET C2M MOSFET Technology. Durham, USA: Cree, Inc.; 2021. Available from: www.cree.com/power

[33] Dumollard Y, Batista E, Pecastaing L, Dienot JM. Modeling of the electrothermal behavior of a SiC MOSFET Chip during a short-circuit and adaptation to an entire power module. In: 8th Euro-Asian Pulsed Power Conference, In Abstract Book of the EAPPC-BEAMS-MEGAGAUSS'2020. Biarritz, France: EAPPC Comittee; 2021. p. 117

## Chapter 3

# Self-Heating Effects in Nanoscale MOSFETs

*Ahmed Yusupov, Atabek Atamuratov and Makhkam Khalilloev*

## Abstract

The ongoing miniaturization of metal-oxide-semiconductor field-effect transistors (MOSFETs) remains a central focus in the advancement of nanoelectronic technologies. As device dimensions continue to shrink into the nanometer scale, various reliability and degradation phenomena become increasingly significant. This chapter addresses one such critical issue—the self-heating effect (SHE)—in nanoscale MOSFETs, with a particular focus on FinFET and junctionless (JL) FinFET architectures fabricated on silicon-on-insulator (SOI) structures. The analysis explores the impact of gate oxide and back oxide material compositions on the magnitude of the SHE. A comparative study is conducted to assess how different oxide materials influence the thermal behavior of JL FinFETs across various channel geometries. Furthermore, the effect of source/drain doping profiles—including both uniform and analytical distributions—as well as channel doping levels, on the SHE in FinFET devices is systematically examined. The chapter provides a comprehensive evaluation of thermal characteristics, ON/OFF current ratio ($I_{on}/I_{off}$), and threshold voltage ($V_{th}$) under varying doping schemes. The findings culminate in a set of conclusions and design guidelines aimed at mitigating self-heating in next-generation nanoscale MOSFET technologies.

**Keywords:** FinFET, Junctioless FinFET, self-heating effect, gate oxide, back oxide, source area doping profile, drain area doping profile, channel doping level

## 1. Introduction

One of the significant trends in contemporary electronics is the scaling of metal-oxide-semiconductor field-effect transistor (MOSFETs), driven by the need to reduce power consumption and increase integration density. As MOSFET dimensions approach the nanometer regime, various degradation effects emerge. Consequently, research efforts increasingly focus on investigating the degradation mechanisms induced by aggressive scaling. In this context, novel device architectures and advanced functional materials are proposed to overcome scaling-related limitations. To mitigate the short-channel effects (SCEs) inherent in traditional planar MOSFETs [1], the FinFET architecture was introduced as an effective solution [2, 3]. Subsequently, technologically simpler junctionless FinFETs were developed [4, 5], also demonstrating enhanced resistance to SCEs.

In both FinFET and junctionless FinFET designs built on SOI platforms, the transistor channel lies directly atop a back-oxide (BOX) layer. Because BOX materials conduct heat much less effectively than bulk silicon, they promote localized heat buildup in the channel—a phenomenon known as the self-heating effect (SHE) [6, 7]. This trapped heat can alter the electrical response and degrade overall device performance.

Self-heating in FETs is influenced by a broad spectrum of factors, including the material and composition of the gate oxide and BOX [8], their geometric configuration [9], the overall device architecture [10], channel doping concentration, and source/drain doping profiles [11]. The following sections provide a detailed analysis and discussion of how these parameters affect the onset and severity of the SHE.

## 2. Material and structural dependencies of SHE

### 2.1 Impact of gate and back oxide layer materials

Oxide selection plays a pivotal role in governing self-heating in FinFETs. Here, we evaluate three common dielectrics—silicon dioxide ($SiO_2$), hafnium dioxide ($HfO_2$), and silicon nitride ($Si_3N_4$)—both as the gate insulator and as the back-oxide layer. Our modeled device is a junctionless FinFET on silicon, as illustrated in **Figure 1**, in accordance with [12], with a uniform n-doped channel and titanium nitride (TiN) gate electrode. Device dimensions and channel doping details appear in **Table 1**.

We carried out all device simulations in Synopsys Sentaurus TCAD. To model self-heating faithfully, we employed a thermodynamic transport framework augmented by a density-gradient term to account for quantum confinement effects. Our mobility model captured both doping-level–dependent scattering and velocity saturation under strong electric fields. For cases using high-κ gate dielectrics (e.g., $HfO_2$), we further included degradation contributions from Coulomb scattering and phonon interactions at the oxide–semiconductor interface.

In these studies, we varied both gate-oxide and back-oxide thicknesses across different material stacks, recording the lattice temperature at the channel's midpoint to quantify self-heating under each configuration.

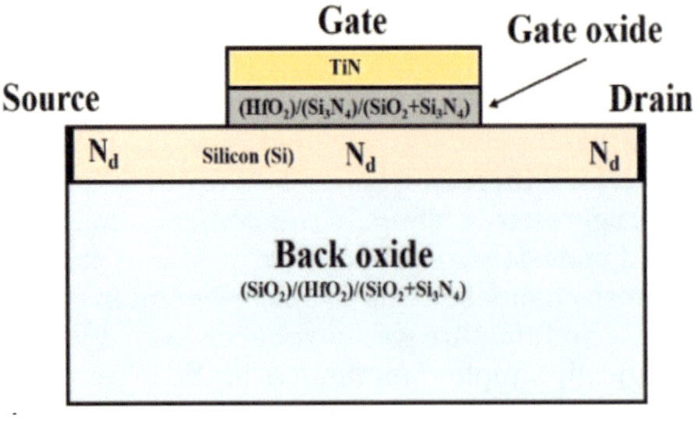


**Figure 1.**
*Cross-section of the simulated JL FinFET [12].*

| Parameter | Designation | Value of the parameter |
|---|---|---|
| Doping level in the channel | $N_d$ | $5 \cdot 10^{18}\ cm^{-3}$ (n-tip) |
| Doping level in the source and drain areas | $N_d$ | $5 \cdot 10^{18}\ cm^{-3}$ (n-tip) |
| Thickness of the Gate oxide ($HfO_2$, $Si_3N_4$, $SiO_2 + Si_3N_4$) | $t_{ox}$ | $t_{EOT}$ = 1.0–1.5 nm |
| Channel thickness | $T_{si}$ | 9 nm |
| Channel width | $W_b$ | 22 nm |
| Back oxide layer ($SiO_2$, $HfO_2$, $SiO_2 + Si_3N_4$) thickness | $T_{box}$ | 10–1000 nm |
| Gate length | $L_{gate}$ | 10 nm |

**Table 1.**
*Geometrical and physical parameters of the considered transistor.*

### *2.1.1 Lattice temperature variation with gate oxide thickness*

**Figure 2** plots how the channel's lattice temperature varies with gate-oxide thickness for different back-oxide materials. Note that we excluded $SiO_2$ as a gate dielectric in this study because at the minimal thickness of 1 nm, its leakage current becomes unacceptably high due to significant tunneling through the oxide layer.

As seen in **Figure 2**, the lattice temperature generally decreases gradually—and predominantly monotonically—with increasing gate-oxide thickness for all materials examined. The temperature drops observed at specific thicknesses correspond to reductions in drain current, which, in turn, lower Joule heating and thus reduce the channel temperature (see **Figure 3**).

Enhanced heat evacuation via the gate oxide also helps lower channel temperatures. While making the gate oxide thicker might seem to boost thermal resistance—thus trapping more heat—the trade-off is a larger interface between the oxide and the gate electrode, which spreads heat more effectively away from the channel. As illustrated in **Figures 3** and **4**, this interface area increases nearly linearly with oxide thickness, directly driving the steady decline in lattice temperature measured at the channel's core.

Moreover, **Figure 2** indicates that the back-oxide material exerts a more pronounced effect on the channel's lattice temperature. This thermal influence is analyzed in detail in the following section.

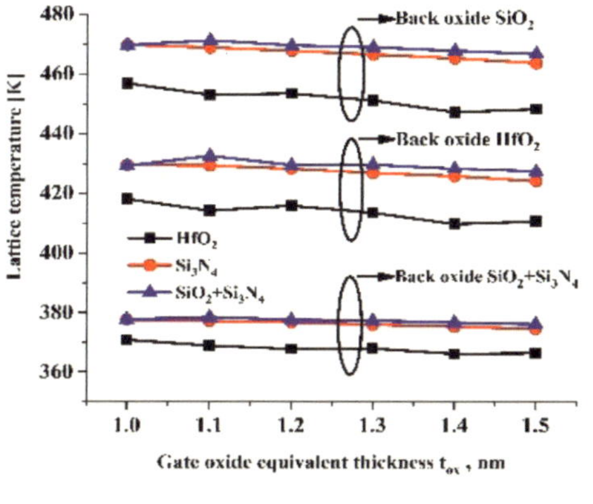


**Figure 2.**
*Lattice temperature variation with equivalent gate oxide thickness for various gate and back oxide materials [12].*

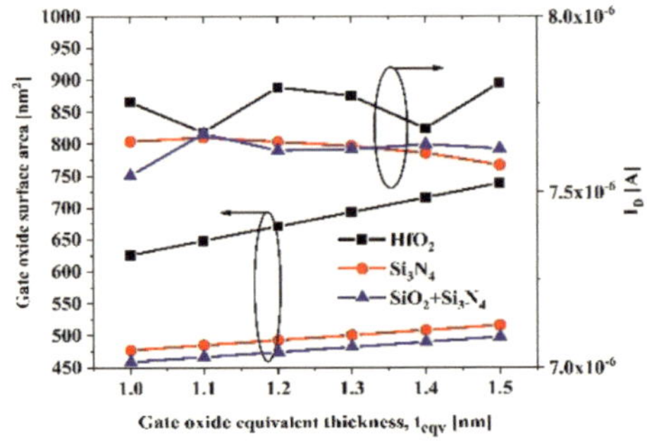


**Figure 3.**
*Effect of equivalent gate oxide thickness on drain current and gate–oxide contact area for different oxide materials [12].*

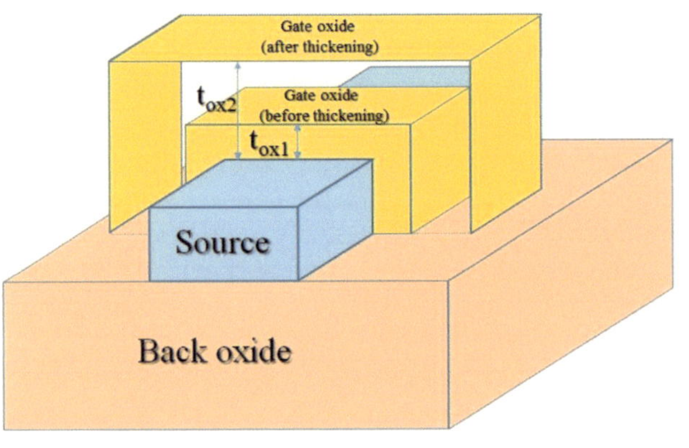


**Figure 4.**
*FinFET structure demonstrating an increase in the contact area between the gate oxide and the gate with increasing gate oxide thickness.*

### *2.1.2 Effect of back oxide thickness on lattice temperature*

**Figure 5** illustrates the dependence of the channel's lattice temperature on the back-oxide thickness for two gate-dielectric configurations: $HfO_2$ and a bilayer $SiO_2$ + $Si_3N_4$ stack. As noted previously, pure $SiO_2$ is not considered here as a standalone gate dielectric due to its high leakage current at the minimum studied thickness of 1.2 nm. At such thin layers, quantum tunneling through the $SiO_2$ becomes significant and may compromise device reliability.

As shown in **Figure 5**, regardless of the gate-dielectric material, increasing the BOX thickness always leads to a rise in lattice temperature. This trend agrees with theoretical models describing the relationship between lattice temperature and BOX thickness in MOS structures [13]. According to these models, the increased thermal resistance caused by a thicker BOX layer reduces heat-removal efficiency, resulting in a higher channel temperature. This behavior is quantitatively described by Eq. (1) derived from the referenced model:

$$\Delta T = \frac{(P_t \cdot T_{ox})}{K_b \cdot A} \tag{1}$$

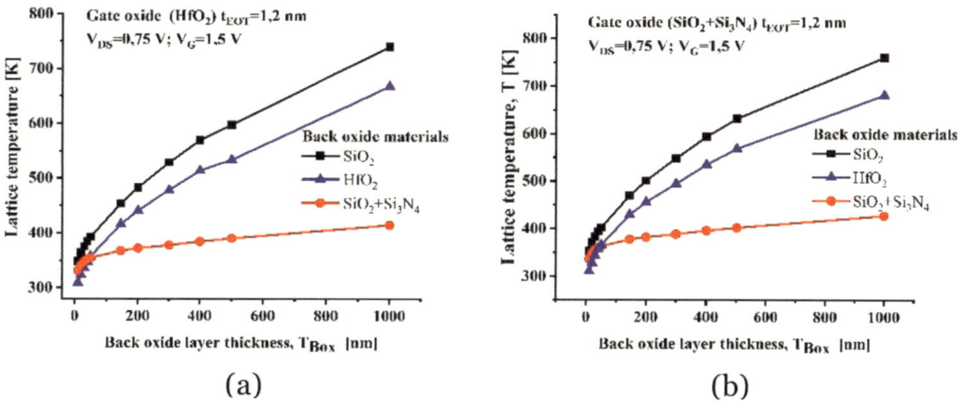


**Figure 5.**
*Lattice temperature variation with BOX thickness for $HfO_2$ (a) and $SiO_2$ + $Si_3N_4$ (b) Gate oxide configurations [12].*

In the context of Eq. (1), $P_t$ represents the power converted to heat by the channel current, $K_b$ is the thermal conductivity of the back-oxide layer, and $A$ is the contact area between the silicon channel and that oxide. When the back-oxide thickness is held constant, using pure $SiO_2$ yields the highest channel temperature, whereas the $SiO_2$ + $Si_3N_4$ composite achieves the lowest. This outcome arises from the trade-off between how much heat is generated-driven chiefly by the drain current-and how effectively that heat is removed, which, in turn, is governed by the oxide's thermal conductivity. **Table 2** lists the thermal conductivities of the considered BOX materials alongside their corresponding drain currents under identical bias conditions. Although the $SiO_2$ + $Si_3N_4$ configuration is a serial multilayer stack, **Table 2** includes only $Si_3N_4$'s thermal conductivity because the $SiO_2$ fraction in the stack is minimal (0.2–20% of the total thickness), and thus its effect on overall thermal resistance is negligible.

As shown in **Table 2**, $SiO_2$ has the lowest thermal conductivity among all BOX materials and, consequently, yields the lowest drain current. Comparing pure $SiO_2$ to the $SiO_2$ + $Si_3N_4$ stack as the BOX, the latter provides a 25% higher drain current and generates 57% more Joule heat. Nevertheless, the effective thermal conductivity of the $SiO_2$ + $Si_3N_4$ structure is roughly 13 times that of pure $SiO_2$. As a result, despite increased heat generation, the much greater heat-dissipation capability of the $SiO_2$ + $Si_3N_4$ stack leads to the lowest channel temperature, whereas pure $SiO_2$ causes the largest temperature rise in the channel region.

| Back oxide materials | Thermal conductivity $K_b$ [W/m K] | Drain current (gate oxide is HfO2, $t_{box}$ = 500 nm, Vg = 1.5 V, Vd = 0.75 V), A |
|---|---|---|
| $HfO_2$ | 2,3 | 787823E-6 |
| $SiO_2$ | 1,4 | 700567E-6 |
| $Si_3N_4$ ($SiO_2$ + $Si_3N_4$) | 18,5 | 87783E-6 |

**Table 2.**
*Thermal conductivity of BOX materials and corresponding drain current.*

Our simulations demonstrate that while both the gate dielectric and the back oxide influence the FinFET's central-channel temperature, the back-oxide layer governs the outcome most strongly because of its thermal conduction properties. The gate oxide does affect drain current (and hence the total heat produced), but the BOX's ability to spread and remove that heat ultimately sets the channel's steady-state temperature.

Across all of the oxide stack combinations we tested, the configuration delivering the lowest self-heating comprises $HfO_2$ as the gate dielectric paired with a $SiO_2 + Si_3N_4$ multilayer serving as the back oxide.

## 2.2 Comparison of gate and back oxide contributions to the self-heating effect in nanoscale junctionless FinFETs with varying channel geometries

A review of the existing scientific literature on the SHE in FinFETs—particularly their junctionless variant (JL FinFET)—reveals that the influence of channel geometry on this phenomenon has received limited attention. This gap is critically important, given that nanometer-scale JL FinFET structures are likely to exhibit channel-shape variations due to process constraints and design flexibility. To address this shortcoming, the present study undertakes three-dimensional modeling and a comparative analysis of how gate-oxide and BOX layers affect self-heating in nanometer-scale JL FinFETs with different channel shapes.

We designed, in accordance with [14], two sets of junctionless FinFETs—each preserving identical channel cross-sectional area—to isolate how gate and back oxides influence self-heating:

- Rectangular-section devices (**Figure 6**): Here, the channel's thickness ($T_{si}$) and width are varied inversely so that their product (the cross-sectional area) remains fixed, yielding purely rectangular profiles.

- Shape-variable devices (**Figure 7**): These transistors keep both the area and the channel's base width ($W_{si}$) constant but morph the cross section from a rectangle into a trapezoid. The defining metric in this group is the $T_{si}/W_{si}$ ratio, which captures the changing channel shape.

The structures shown in **Figures 6** and 7 were modeled and analyzed in three dimensions using the Synopsys Sentaurus TCAD suite [15]. To improve accuracy at

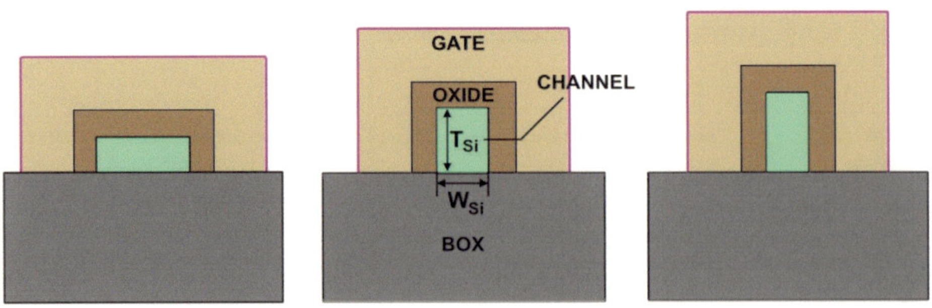


**Figure 6.**
*Schematic depiction of JL FinFET structures with a rectangular channel cross section: devices featuring different width-to-thickness ratios while maintaining a constant cross-sectional area [14].*

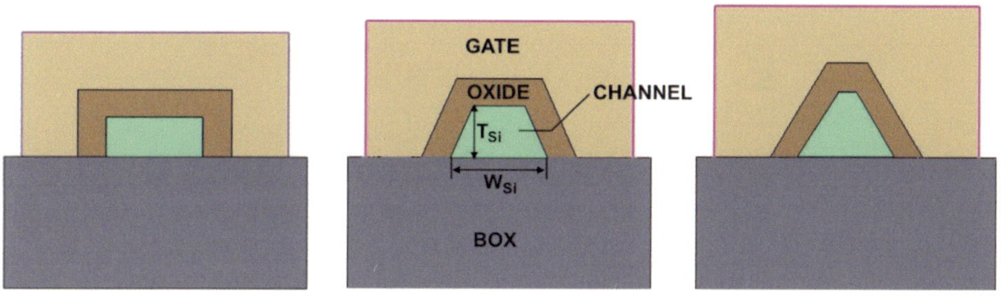


**Figure 7.**
*Modeled JL FinFET structures with rectangular and trapezoidal channel shapes: constant cross-sectional area and base width, with variable channel thickness [14].*

the nanoscale, standard physical models were augmented with mobility-degradation mechanisms that depend on doping concentration.

Our simulations incorporated a comprehensive set of physical phenomena to capture both charge transport and thermal effects at the nanoscale. We modeled impurity scattering and carrier-velocity saturation under strong electric fields, and included the normal field component to accurately represent boundary-region impacts on drain current. Self-heating was addressed via a thermodynamic transport formulation coupled with Shockley–Read–Hall recombination dynamics.

To account for mobility degradation in the presence of a high-κ gate dielectric, we adopted the Lombardi mobility model with empirical terms for remote Coulomb interactions and phonon scattering at the oxide–semiconductor interface. Quantum confinement and tunneling influences were represented through a density-gradient correction integrated into the combined drift-diffusion and thermodynamic transport framework [16].

The modeling approach was calibrated against experimental data from [17]. The technology parameters used—corresponding to a 10 nm CMOS process—are listed in **Table 3**. The resulting simulated and experimental transfer characteristics are shown in **Figure 8**.

Joule heating, driven by the channel current, is the main source of temperature rise in the transistor. That channel current breaks down into three parts: the drift component ($J_{dr}$), the diffusion component ($J_{dif}$), and the thermoelectric component ($J_t$). To explore how device geometry and material properties affect self-heating, we

| Parameters | Designation | Values |
|---|---|---|
| Channel doping level | $N_{sub}$ | $5.10^{18} cm^{-3}$ (n-type) |
| Thickness of $HfO_2$ gate oxide layer | $t_{ox}$ | 6,7 nm ($t_{eqv}$ = 1,2 nm) |
| Thickness of the channel | $T_{si}$ | 9–22 nm |
| Width of the channel base | $W_b$ | 22–5 nm |
| The thickness of the $SiO_2$ back oxide layer | $T_{box}$ | 145 nm |
| Width of the $SiO_2$ back oxide layer | $W_{box}$ | 69.4 nm |
| TiN Gate length | $L_{gate}$ | 10 nm |

**Table 3.**
*Parameters used in device simulations.*

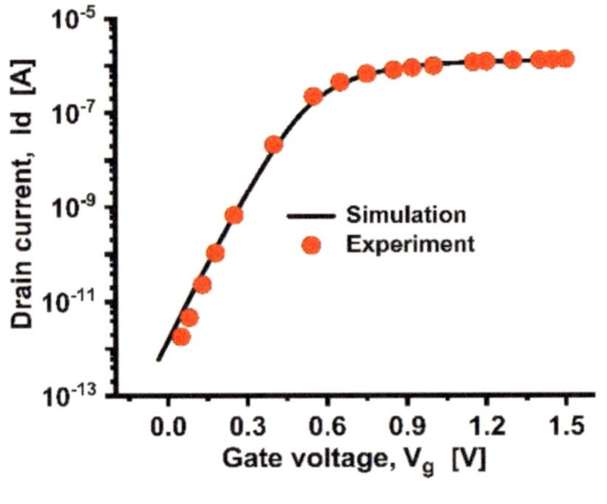


**Figure 8.**
*Comparison of simulated and measured transfer curves for the transistors in [17], using the same device dimensions and bias conditions.*

conducted an in-depth study of nanometer-scale JL FinFET designs from both device groups outlined earlier.

### *2.2.1 Performance evaluation of a nanoscale junctionless FinFET with rectangular channel geometry*

**Figure 9(a)** depicts the current density measured at the middle of the channel for different $T_{Si}/W_{Si}$ ratios. In **Figure 9(b)**, you can see how the modeled electric field strength changes as $T_{Si}/W_{Si}$ varies. **Figure 9(c)** then maps out the carrier

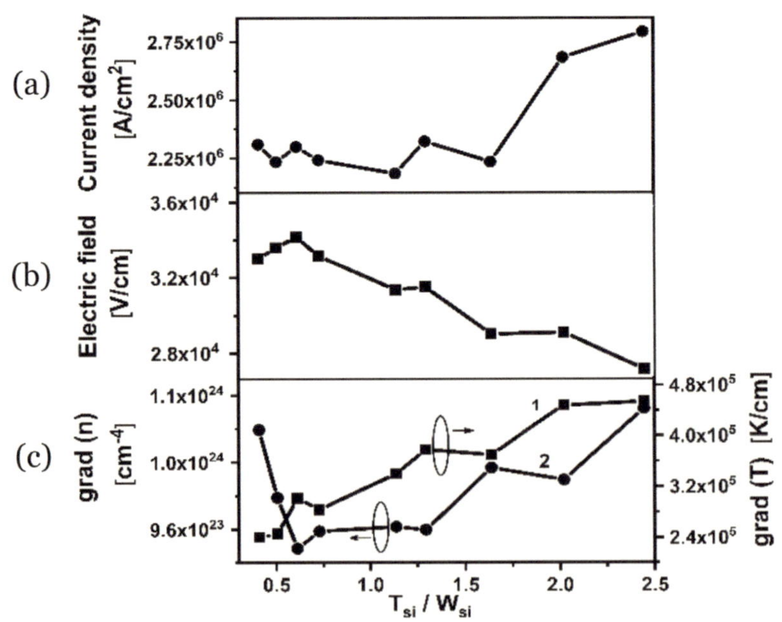


**Figure 9.**
*Variation of (a) current density, (b) electric field, and (c) carrier concentration (curve 1) and temperature gradients (curve 2) with $T_{Si}/W_{Si}$ at the channel center [14].*

concentration and temperature gradient profiles along the same axis for each $T_{Si}/W_{Si}$ value. All these data come from sampling the electric field, carrier density, and temperature along the channel's centerline.

Typical spatial distributions of these quantities—for example, for $T_{Si}/W_{Si}$ = 0.4—are shown in **Figure 10**. Based on the profiles in **Figure 9**, all components of the current density were quantitatively evaluated. For instance, at $T_{Si}/W_{Si}$ = 1.3, the drift current density ($j_E$), the diffusion current density ($j_{\Delta n}$), and the thermoelectric current density ($j_{\Delta T}$) were calculated using Eqs. (2)–(4).

$$j_E = en\mu E \tag{2}$$

$$j_{\Delta n} = eD\frac{\Delta n}{\Delta x} \tag{3}$$

$$j_{\Delta T} = \sigma S\frac{\Delta T}{\Delta x} \tag{4}$$

where:

e is the elementary charge, n is the carrier concentration, μ is the electron mobility, D is the electron diffusion coefficient in silicon, σ is the electrical conductivity of silicon, S is the Seebeck coefficient for silicon.

The calculation results yielded the following values:

$$j_E = 6.98\times10^6 \text{ A}/cm^2,$$

$$j_{\text{Än}} = 4.83\times10^6 \text{ A}/cm^2,$$

$$j_{\Delta T} = 3.78\times10^2 \text{ A}/cm^2.$$

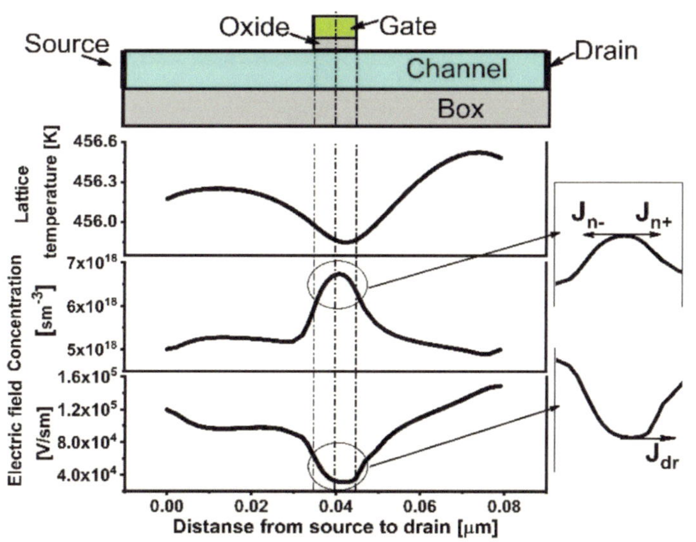


**Figure 10.**
*Spatial profiles of temperature, carrier concentration, and electric field along the channel's central axis for $T_{Si}/W_{Si}$ = 0.4 [14].*

Evidently, the thermoelectric current density is several orders of magnitude smaller than the drift and diffusion components. Therefore, one can conclude that the total current density is dominated by the drift and diffusion currents, whose magnitudes are comparable (as illustrated by the lower and upper insets in **Figure 10**).

**Figure 10** also shows that the carrier-concentration gradient in the channel consists of two components pointing in opposite directions and of different magnitudes. This asymmetry arises from the nonuniform influence of the vertical electric field—induced by the gate—at the source and drain ends. Specifically, one diffusion component ($Jn^{+}$) is directed along the source-to-drain electric field, while the other ($Jn^{-}$) flows oppositely (see the upper inset of **Figure 10**). Thus, the net diffusion current, defined as $Jn = Jn^{+} - Jn^{-}$, opposes the drift current. As a result, the overall channel current is determined by the difference between the drift and net diffusion currents.

**Figure 9** shows that as the $T_{Si}/W_{Si}$ ratio grows—especially beyond about 1.75—the diffusion-driven portion of the current takes on a larger share of the total current density. Yet, **Figure 11** reveals that the channel's lattice temperature climbs steadily with $T_{Si}/W_{Si}$ across the full range, a trend that does not map directly onto the current-density behavior.

This mismatch tells us that simply looking at how much current flows is not enough to predict temperature changes. Instead, the ability of the transistor stack to move heat away from the channel—via both the back-oxide and gate-oxide layers—plays the decisive role. In fact, one can use the heat-conduction relationship given in Eq. (1) to capture how the channel temperature scales with BOX thickness, as derived in [13].

To compare heat conduction through the back-oxide and gate-oxide layers, Eq. (1)—which applies equally to both layers since their heat-transfer mechanisms are identical—was used. The cross-sectional area for the back oxide is

$$A = W_{Si} \times L,$$

and for the gate oxide

$$A = (2T_{Si} + W_{Si}) \times L,$$

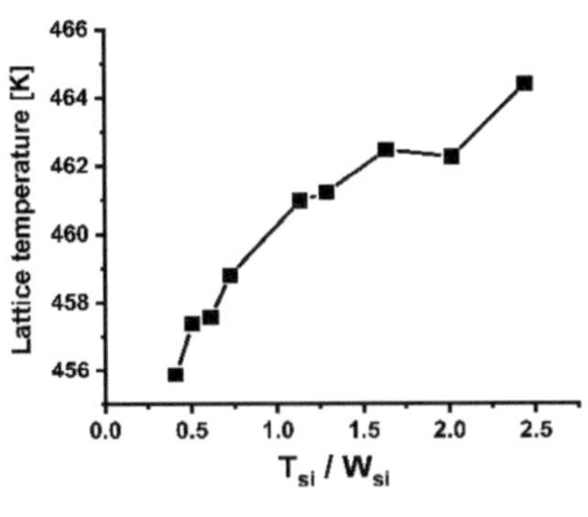


**Figure 11.**
*Dependence of the temperature in the middle of the channel on the ratio $T_{Si}/W_{Si}$ for transistors of the first group [14].*

where $L$ is the channel length. Thus, the temperature rise normalized to the dissipated power can be expressed by Eqs. (5) and (6):

$$\frac{\Delta T}{P_t} = \frac{T_{box}}{K_{box} \cdot W_{Si} \cdot L} \quad (5)$$

$$\frac{\Delta T}{P_t} = \frac{T_{gox}}{K_{gox} \cdot (2 \cdot T_{Si} + W_{Si}) \cdot L} \quad (6)$$

In these expressions:

$T_{box}$ is the back-oxide thickness,

$T_{gox}$ is the gate-oxide thickness,

$K_{box}$ and $K_{gox}$ are the thermal conductivities of the back-oxide and gate-oxide layers, respectively.

In our simulations, $SiO_2$ served as the back oxide and $HfO_2$ as the gate dielectric. We adopted thermal conductivities of 1.5 W/(m·K) for $SiO_2$ and 0.6 W/(m·K) for $HfO_2$, based on their respective ranges (1.2–2 W/(m·K) for $SiO_2$ over 200–1500 K, and 0.49–0.95 W/(m·K) for $HfO_2$ over 300–500 K). **Figure 12** demonstrates that this contrast in conductivities makes the back oxide far more effective at channel cooling: the $\Delta T/P_t$ curve for the back-oxide route (curve 1) climbs steadily with $T_{Si}/W_{Si}$, matching the temperature rise observed in **Figure 11**. By comparison, when heat flows out through the gate oxide, $\Delta T/P_t$ actually falls as $T_{Si}/W_{Si}$ increases (see inset of **Figure 12**). These results underscore the preeminent role of the back oxide layer in extracting heat and regulating the channel's core temperature.

### *2.2.2 Performance analysis of nanoscale JL FinFET with different channel shapes*

We evaluated how gate-oxide versus back-oxide layers contribute to self-heating by running temperature simulations for devices whose channel cross-section morphs from a rectangle to a trapezoid (see **Figure 7**), while holding both channel width and area constant. As a result, the thickness-to-width ratio $T_{Si}/W_{Si}$ only spans 0.4 to 0.7. The data in **Figure 13** reveal that, despite altering the channel's shape, the central

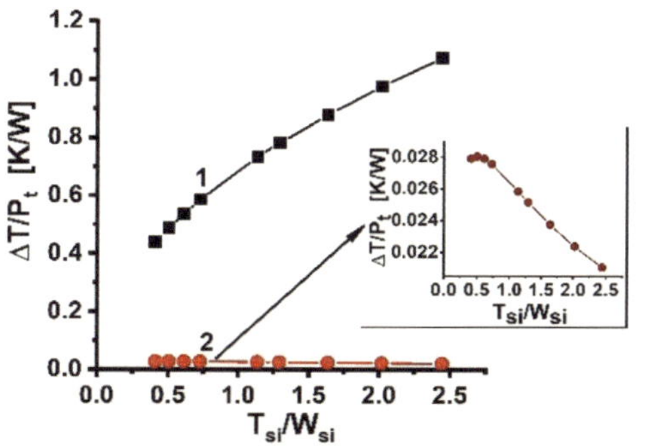


**Figure 12.**
*$\Delta T/P_t$ dependence on $T_{Si}/W_{Si}$ in cases of the heat dissipation through back oxide (curve 1) and gate oxide (curve 2) layers for the transistors of the first group [14].*

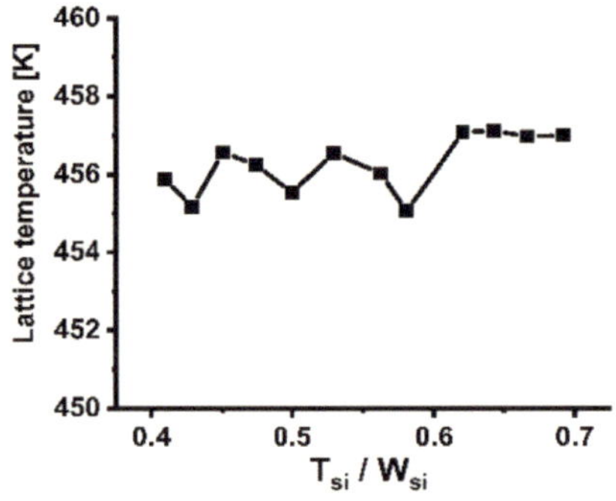


**Figure 13.**
*Dependence of the temperature in the middle of the channel on the ratio $T_{Si}/W_{Si}$ for transistors of the second group [14].*

temperature barely changes—because the base width touching the back oxide remains the same throughout these shape transitions.

When heat is removed via the back-oxide layer, the ratio $\Delta T/P_t$ also shows minimal variation as $T_{Si}/W_{Si}$ increases (see **Figure 14**). In contrast, the dependence of $\Delta T/P_t$ on $T_{Si}/W_{Si}$ for the gate-oxide path is non-monotonic (inset of **Figure 14**), a behavior that does not match the channel-center temperature's insensitivity to cross-section shape. Therefore, even in this case, the back-oxide layer plays the primary role in heat removal.

In nanometer-scale junctionless FinFETs, regardless of how the channel's cross section is shaped, the temperature at the channel's midpoint is controlled primarily by how well the oxide layers conduct heat and how much surface area they share with the silicon. In devices that employ $SiO_2$ beneath the channel and $HfO_2$ above it, the width of the channel's base—where it meets the back oxide—emerges as the critical geometric parameter. When the channel's thickness-to-width ratio grows ($T_{Si}/W_{Si}$ rises) without changing its overall cross-sectional area, this base becomes narrower, causing the central temperature to climb. Therefore, to minimize self-heating and boost reliability in these tiny FinFETs, one should design the channel so that its interface with both the gate and back oxides is as large as possible.

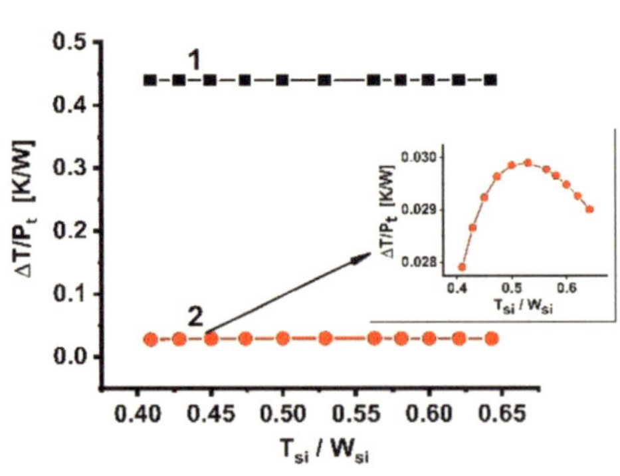


**Figure 14.**
*$\Delta T/P_t$ dependence on $T_{Si}/W_{Si}$ in cases of the heat dissipation through the back oxide (curve 1) and gate oxide (curve 2, inset) layers for the transistors of the second group [14].*

## 2.3 Impact of source (drain) doping profiles and channel doping level on self-heating effect in FinFET

A wealth of research has investigated self-heating in FETs, often emphasizing how device dimensions [18], choice of materials [19, 20], and architectural layouts [18, 21] influence the phenomenon. Yet, how source/drain implant profiles and overall channel doping levels affect thermal buildup has received far less attention, even though these factors are deeply tied to fabrication processes. Carefully engineering the dopant gradients at the source and drain—and tuning the channel's doping concentration—offers a promising route to bolster a transistor's resistance to self-heating.

To investigate this important practical aspect, the present work studies how the channel-center temperature depends on source/drain doping profiles and channel doping level in nanoscale FinFETs. The device structure and key simulation parameters are shown in **Figure 15** and **Table 4**, respectively. In particular, the analysis explores how the doping profile shape influences self-heating, the on/off current ratio ($I_{on}/I_{off}$), and the threshold voltage ($V_t$).

Two distinct types of source/drain doping profiles are considered. The corresponding transistor structures—with color-coded impurity concentrations—are

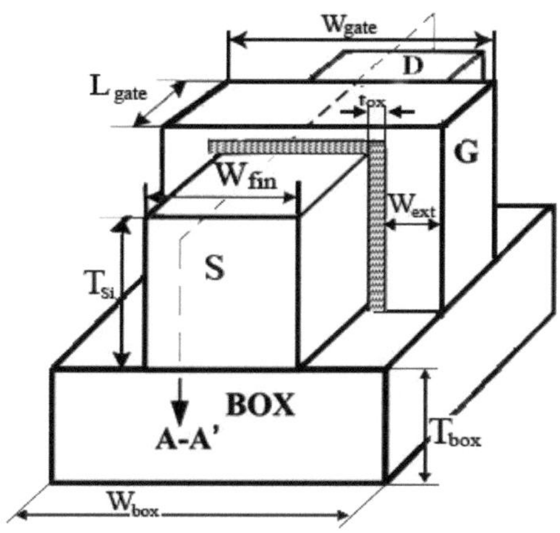


**Figure 15.**
*The structure of the simulated SOI FinFET.*

| Parameter | Designation | Value |
|---|---|---|
| Channel doping level | $N_a$ | $10^{15}$–$10^{17}$ cm$^{-3}$ |
| Maximal doping level in source and drain areas | $N_d$ | 5.10$^{18}$ $cm^{-3}$ |
| Thickness of the gate oxide ($HfO_2$) | $t_{ox}$ | 6,7 nm ($t_{eqv}$ = 1,2 nm) |
| Thickness of the channel | $T_{si}$ | 9 nm |
| Width of the channel | $W_b$ | 22 nm |
| Thickness of the back oxide | $T_{box}$ | 145 nm |
| Gate length | $L_{gate}$ | 10 nm |

**Table 4.**
*Parameters of the simulated FinFET.*

shown in **Figure 16(a)** and **(b)**. The first structure features a uniform (constant) doping profile (**Figure 16(a)**), while the second employs an analytical (Gaussian) profile (**Figure 16(b)**). The spatial distributions of doping concentration along the channel—from source to drain—for both profiles appear in **Figure 17**.

For the analytical profile, two scenarios were examined, differing in impurity concentration at the channel–source/drain interfaces. For clarity, devices were classified into three categories based on their doping profile, as schematically illustrated in **Figure 18**. The first category includes transistors with a uniform profile (**Figure 17(a)**). The second comprises devices with a Gaussian profile that maintains an interface concentration of $3 \times 10^{17}$ $cm^{-3}$ at the channel–source/drain boundaries (**Figure 17(b)**). The third category consists of transistors whose Gaussian profile has its boundary concentration equal to the channel's doping level (**Figure 17(c)**).

Simulations were performed using the Synopsys Sentaurus TCAD suite. Accurately capturing self-heating in FinFETs requires the inclusion of electrothermal effects in the channel alongside the conventional drift-diffusion transport mechanism. The thermodynamic transport model in Sentaurus accounts for both drift-diffusion–driven current and for current induced by electrothermal effects. For comparison, Id–Vg characteristics were computed using only the drift-diffusion model and then with electrothermal effects for the uniform-profile device (**Figure 19**). Clearly, including electrothermal effects produces a markedly different Id–Vg curve. In this study, all calculations employed the thermodynamic transport model. Density-gradient corrections were also applied to account for quantum effects, and the mobility model included doping-dependent degradation and velocity saturation under high electric fields.

To calibrate the simulation model, the computed Id–Vg characteristics were compared with experimental data reported in (**Figure 20**) [22]. The experimental transistor had a gate length of 25 nm, so we first simulated a device with the same gate length to verify the accuracy of our model. The calibrated model was then applied to a transistor with a reduced gate length of 10 nm.

To assess the impact of gate length on channel temperature, we calculated the temperature difference between the drain-end region and the channel center for the 25 nm device using Fourier's law of heat conduction [23]. We used a silicon-channel thermal conductivity of 149 W/(m·K) and determined the dissipated power at gate bias Vg = 1 V and drain bias Vd = 0.75 V. The calculations yielded a temperature

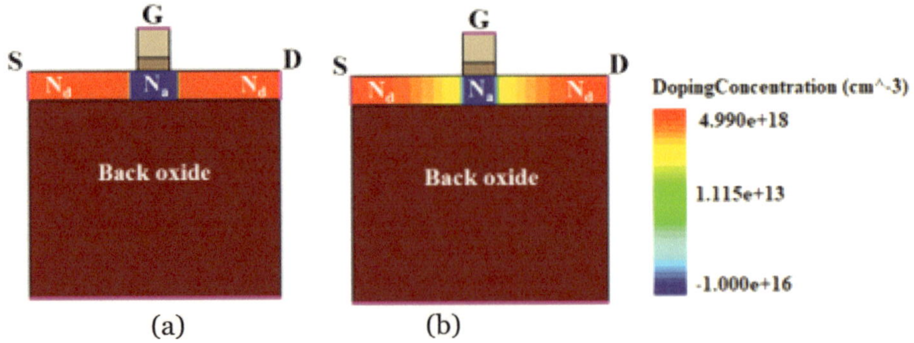


**Figure 16.**
*Cross-sectional views of the modeled FinFETs with (a) a uniform source/drain doping profile and (b) an analytical (Gaussian) doping profile [11].*

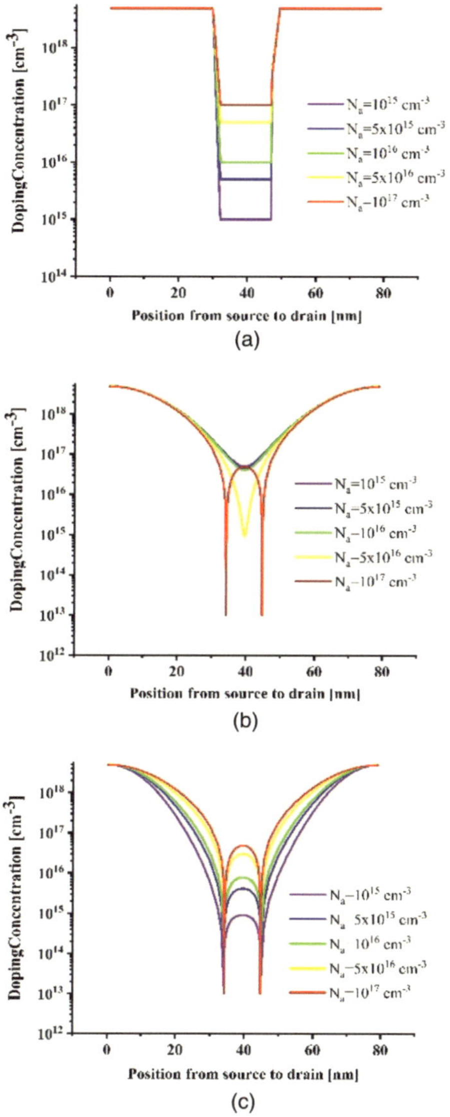


**Figure 17.**
*Distributions of the doping concentrations along the channel from source to drain for the constant profile (a), for the analytical profile with constant edge value 1017 cm$^{-3}$ near source (drain)-channel border (b) and analytical profile with edge value equal to channel doping level (c) [11].*

difference of 0.5 K—approximately 0.2% at 300 K—indicating that the calibrated model can be used for the devices under study with minimal error.

The simulated temperature profile along the channel shows a nonuniform distribution with a pronounced hot-spot near the drain (**Figure 21**). This effect arises because the kinetic energy of accelerating electrons is higher near the drain; consequently, their collisions with the crystal lattice release more heat in that region.

For greater precision in monitoring and given the small temperature difference between the drain region and the channel center, self-heating was evaluated based on the channel-center temperature extracted from the simulated profile. Electrode temperatures were fixed at 300 K in all simulations.

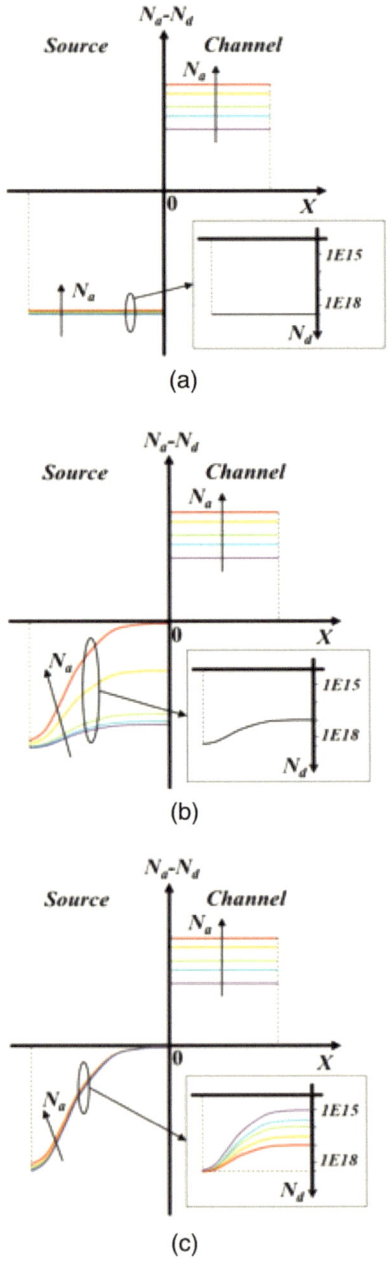


**Figure 18.**
*Schematic representation of the doping profiles for cases at stake: (a) - constant doping profile, (b)- analytical doping profile with constant concentration at the edge of the profile, (c)- analytical doping profile with concentration at the end equal to the doping level in the channel. The insertions display the doping profiles exclusively for the source areas/regions [11].*

Channel temperature depends on the thermal conductivities of both the gate and back oxides. In all modeled devices, $HfO_2$—a high-κ material with a thermal conductivity ranging from 0.49 to 1.25 W/(m·K) over 300–500 K [24]—was used as the gate dielectric. $SiO_2$, with a thermal conductivity between 5 and 11 W/(m·K)

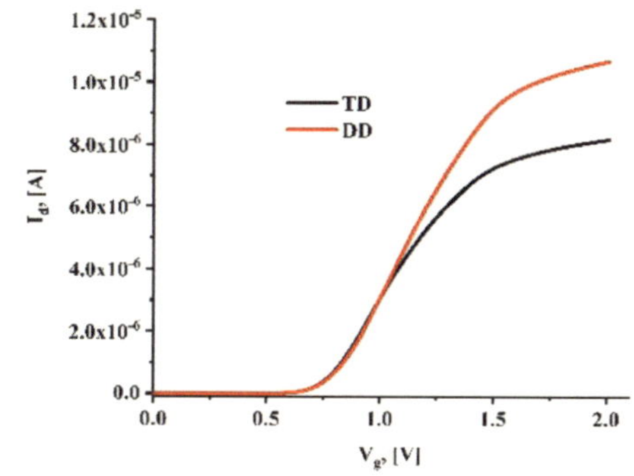


**Figure 19.**
*Simulation of the $I_d$-$V_g$ transfer characteristics for a transistor with constant doping profile using the drift-diffusion (DD) and thermodynamic (TD) transport models. $V_d$ = 750 mV [11].*

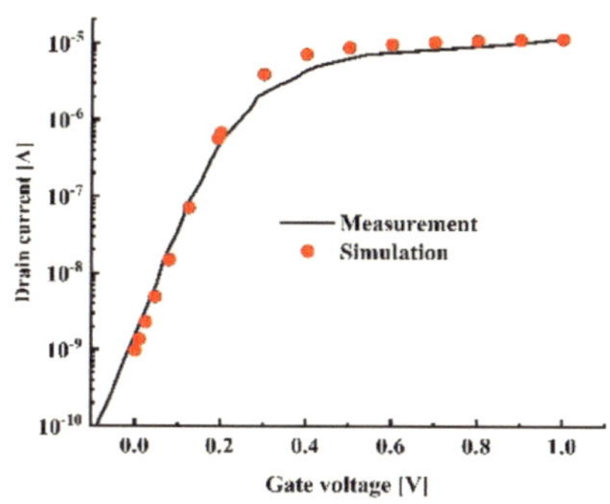


**Figure 20.**
*Comparison of the $I_d$-$V_g$ transfer characteristics of simulated and experimental SOI FinFET transistors for $L_{gate}$ = 25 nm and $V_d$ = 50 mV [11].*

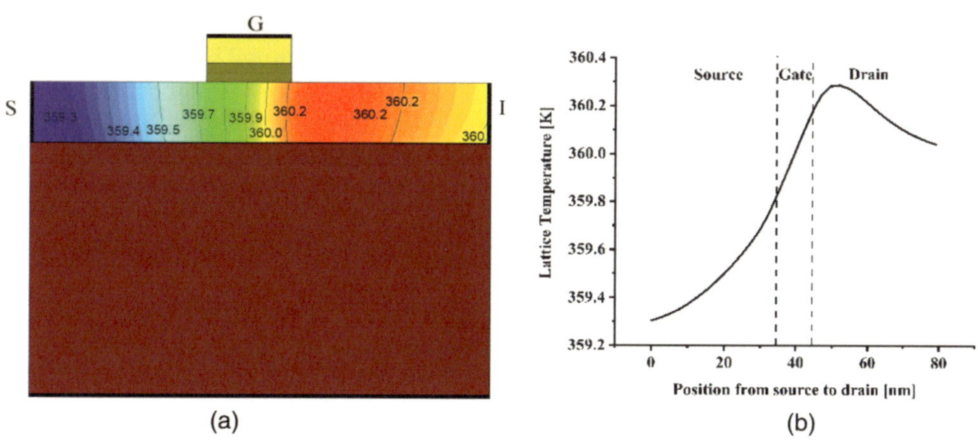


**Figure 21.**
*Distribution of the temperature in the channel (a) and along the channel center (b). $V_g$ = 1 V, $V_d$ = 0.75 V [11].*

over 200–500 K [25], served as the back oxide. However, since this study focuses on the influence of source/drain doping profiles on channel temperature for devices using the same oxide materials, we did not compare material thermal properties or boundary conditions; the heat-removal characteristics remain identical across all cases.

### *2.3.1 Temperature in the center of the channel*

**Figure 8** presents the simulated channel-center temperature as a function of channel-doping level for all three transistor types at $V_{DS}$ = 0.75 V and $V_{GS}$ = 1.0 V.

From the plot, it is evident that the channel-center temperature depends on the source/drain doping profiles. The highest temperature occurs in the first-type devices, while the lowest is in the third type. For all three types, temperature varies monotonically with channel-doping level: in the first and second types, it decreases as doping increases, whereas in the third type it instead increases steadily with higher channel doping.

The average channel-center temperature reflects the balance between heat generation rate and the structure's heat-dissipation efficiency. Heat removal depends on device construction and the surrounding materials, with dielectric thermal conductivities and interfacial thermal resistances playing key roles. In our case, since all transistors share the same structure and materials, temperature differences arise solely from different heat generation in the channel. The primary heat source is Joule heating, which depends on the drain current—determined by the applied biases and carrier concentration in the channel. Because $V_{GS}$ and $V_{DS}$ are identical for all devices, temperature variations stem from differences in current density caused by the various doping profiles. Thus, channel-carrier concentration is the principal factor driving the observed temperature differences.

**Figure 22** shows the simulated electron-concentration versus channel-doping level for all transistor types at $V_{DS}$ = 0.75 V and $V_{GS}$ = 1.0 V. **Figures 22** and **23** demonstrate that changes in temperature closely correlate with changes in channel-center electron concentration, confirming the dominant role of carrier density in shaping the thermal profile.

These results can be interpreted as follows. In n-channel devices with an inverted p-substrate surface, electrons are the majority carriers; in the p-substrate they act as minority carriers. The surface electron density $n_s$ is determined by the surface potential $\Psi_s$ and the equilibrium electron concentration $n_{p_0}$ in the p-substrate via a standard relationship [24]. $\Psi_s$ itself depends on gate voltage $V_g$ and depth from the surface. Treating the channel as the base of a lateral p-n junction (source-channel and drain-channel transitions), the equilibrium electron concentration $n_{p_0}$ in the channel

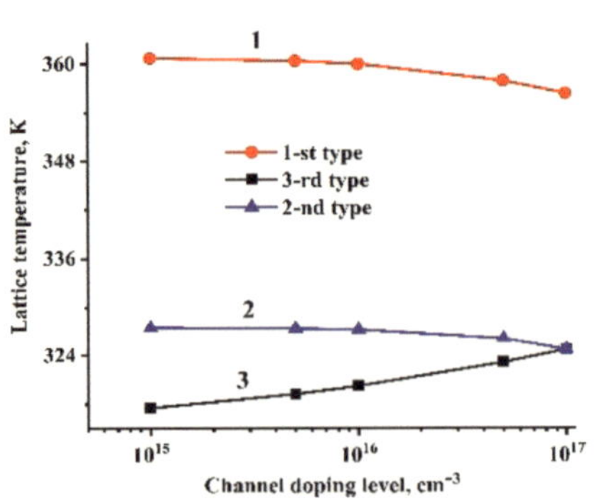


**Figure 22.**
*Lattice temperature dependence on the channel doping level for different transistor types [11].*

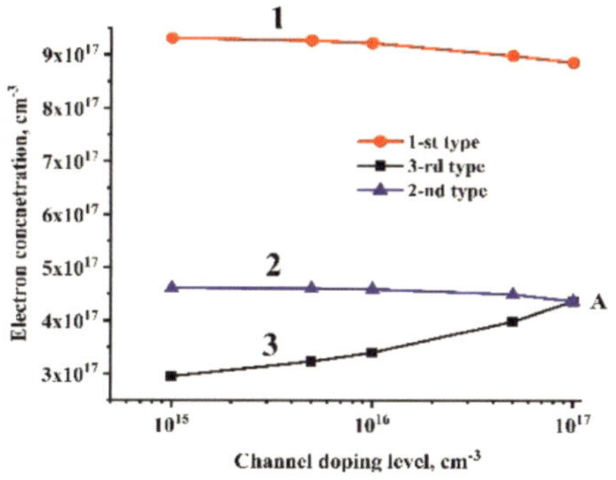


**Figure 23.**
*Dependence of electron concentration at the center of the channel on the channel doping level for various transistor types. $V_{DS}$ = 0.75 V and $V_G$ = 1.0 V [11].*

depends also on the n-region concentration $n_{n_0}$ (source or drain side) and the local potential V [24].

Assuming full ionization of dopants, $n_{n_0}$ equals the source/drain doping level. For an unbiased p-n junction, V equals the contact potential φc. Under applied biases $V_0$ and $V_g$, the local potential $V_0$ is their sum.

In the abrupt p-n case with uniform doping, $n_{n_0}$ equals the source/drain doping.

For the analytical (nonuniform) profile, one defines an effective carrier concentration $n_{n0eff}$—the average dopant concentration near the junctions.

As seen in **Figure 18**, $n_{n0eff}$ is lower in second- and third-type devices than in the first. The third type shows the lowest effective concentration, since its interface doping matches the channel doping level (**Figure 18c**). This lowers the equilibrium electron density, reducing drain current relative to the uniform case, consistent with the concentration simulations of **Figure 23**.

The coincidence of electron concentrations in the channel center for the second and third types at a channel doping of $10^{17}$ $cm^{-3}$ (point A in **Figure 23**) follows from their identical profiles at that level (**Figure 18b** and **c**).

The temperature drop with increasing channel doping in the uniform-profile devices (**Figure 22**) stems from reduced minority-carrier density $n_p$ in the inverted channel, since $n_p{\cdot}p_p = n_0^2$, where $p_p$ is the majority-carrier density and $n_0$ is silicon's intrinsic concentration. Thus, higher $p_p$ lowers $n_p$, cutting drain current and channel temperature. The same trend—falling electron concentration and temperature with more channel doping—holds for the second type.

**Figure 24a** and **b** show that $n_p$ decreases with channel doping in types one and two, via the along-channel electron-density profiles.

In the third-type devices, the channel-carrier concentration is set by both increased channel doping and raised doping near the channel from the source/drain side. Hence, while channel doping reduction lowers electron density, the simultaneous increase at the junctions boosts it, with the latter effect dominating. This behavior is clear in the electron-density *versus* channel-doping plot for type three (**Figure 24c**).

### *2.3.2 $I_{on}/I_{off}$ ratio and threshold voltage*

Considering the impact of source and drain doping profiles on FinFET resilience to self-heating, it is also important—when optimizing device performance—to account for how these profiles affect the on/off current ratio ($I_{on}/I_{off}$) and the threshold voltage ($V_{th}$).

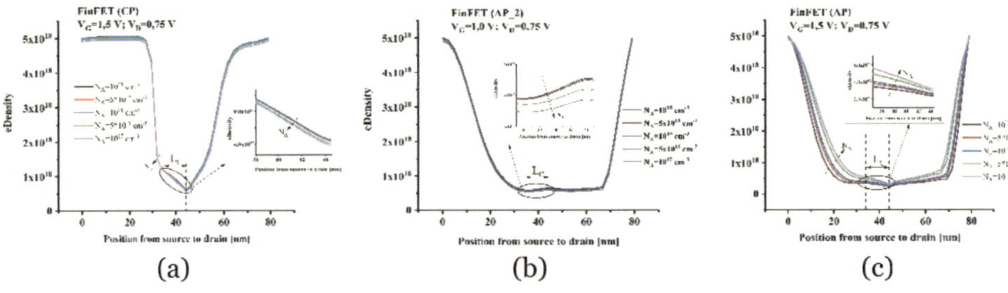


**Figure 24.**
*Distribution of electron concentration along the center of the channel at various channel doping levels for 1-st (a), 2-nd (b) and 3-rd (c) transistor types. $V_{DS}$ = 0.75 V and $V_G$ = 1.0 V [11].*

When the channel doping level increases from $10^{15}$ to $10^{17}$ cm$^{-3}$, $V_{th}$ for the first- and second-type transistors rises monotonically by about 10 mV, whereas for the third-type devices it decreases by roughly 27 mV (**Figure 25a**). This divergence in $V_{th}$ behavior results from differences in carrier concentration at the source/drain regions for each transistor type.

It is well known that a MOSFET's threshold voltage depends on channel-carrier density [26]. However, in our short-channel devices ($L_g$ = 10 nm), $V_{th}$ is also influenced by source/drain carrier concentration. Indeed, when bias is applied to the source–channel or drain–channel junction, the channel's equilibrium electron concentration, $n_{p_0}$, becomes a function of the source/drain electron concentration $n_{n_0}$, as described in Section 2.3.1. Thus, increasing $n_{n_0}$ reduces the channel's hole concentration $p_{p_0}$ and thereby lowers $V_{th}$.

For an approximate comparison of $V_{th}$ across all three transistor types, one may treat the average doping level as $n_{n_0}$ in the analytical-profile devices (types 2 and 3). This approach shows that the highest $n_{n_0}$—and hence the lowest $V_{th}$—occurs in the uniform-profile device (type 1, **Figure 25a**). The lowest average $n_{n_0}$—and therefore the highest $V_{th}$—belongs to the type 3 device. At a channel doping of $10^{17}$ cm$^{-3}$, the source/drain profiles for types 2 and 3 coincide, so their Vth values are equal (**Figure 25a**).

The monotonic $V_{th}$ increase with channel doping in types 1 and 2 (**Figure 25a**) follows the well-known $V_{th}$–doping relationship [26]. In contrast, type 3 devices exhibit

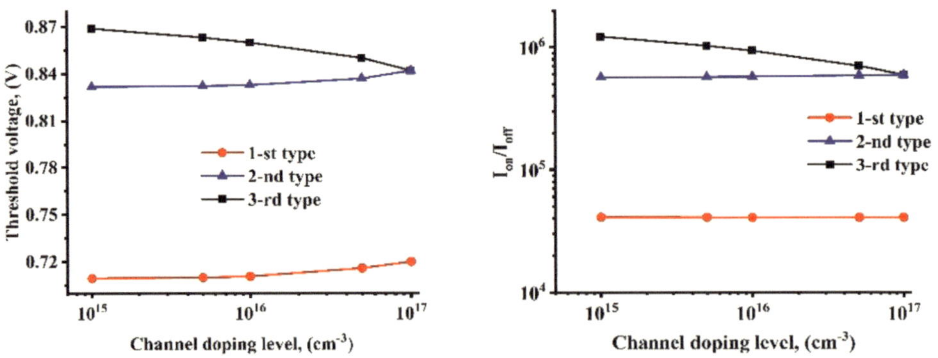


**Figure 25.**
*Threshold voltage (a) and $I_{on}/I_{off}$ ratio (b) dependence on the channel doping level for different transistor types. $V_{DS}$ = 0.75 V [11].*

a $V_{th}$ decrease with higher channel doping. This arises from over-compensation of hole depletion by enhanced electron injection from the source/drain: as channel doping rises, the electron concentration at the profile boundary increases (**Figure 18c**), which dominates and drives $V_{th}$ downward.

**Figure 25b** shows that, at low channel-doping levels ($p_{p_0} < 10^{17}$ $cm^{-3}$), $I_{on}/I_{off}$ is highest for type 3 devices and lowest for type 1. This is attributed to the fact that the average concentration of majority carriers (electrons) in the source and drain regions is significantly lower in the third type, by one to two orders of magnitude, compared to the first and second types, resulting in reduced electron injection into the channel. Consequently, the minority carrier concentration in the channel is reduced, leading to a lower saturation current in the lateral junction, which primarily contributes to the decrease in ($I_{off}$). In this scenario, ($I_{on}$) is also reduced by a factor of 2–3 for the third type due to its higher threshold voltage compared to the first and second types. This would typically suggest a smaller ($I_{on}/I_{off}$) ratio for the third type. However, despite the lower ($I_{on}$), the ($I_{on}/I_{off}$) ratio is actually higher for the third type because ($I_{off}$) is significantly lower—by 1–2 orders of magnitude, than in the first and second types. This difference arises from the variation in the average doping level ($n_{n0}$) in the lateral source (or drain) areas/regions.

In types 1 and 2, $I_{on}/I_{off}$ grows with channel doping (**Figure 25b**) because both $I_{on}$ and $I_{off}$ decrease by similar orders of magnitude: $I_{on}$ drops due to $V_{th}$ rise (**Figure 25a**), and $I_{off}$ diminishes as $n_p$ falls with higher $p_{p_0}$.

For type 3, raising channel doping from $10^{15}$ to $10^{17}$ $cm^{-3}$ halves $I_{on}/I_{off}$ (**Figure 25b**). This reflects the concurrent increase in electron concentration at the source/drain boundaries—which raises $n_p$ in the channel, boosting $I_{off}$—and thus reducing $I_{on}/I_{off}$.

In summary, source/drain doping profiles exert a significant influence on FinFET self-heating by altering channel-carrier concentration. Under identical bias conditions, the largest self-heating occurs in devices with uniform source/drain doping. Employing an analytical (Gaussian) profile reduces channel temperature by about 10% and increases $I_{on}/I_{off}$ by roughly an order of magnitude, while raising $V_{th}$ by approximately 15% compared to the uniform-profile case.

## 3. Conclusions

The results of the study on self-heating in nanoscale MOSFETs indicate the following:

- Both the gate and back oxide selections shape the channel's lattice temperature, with the back oxide's material choice exerting the strongest influence.

- Among the tested dielectrics, pairing a hafnium-oxide gate layer with a silicon-oxide/nitride ($SiO_2$ + $Si_3N_4$) stack beneath the channel delivers the greatest resistance to self-heating.

- In nanometer-scale, junctionless FinFETs—regardless of channel cross-sectional profile—the two dominant factors driving channel-center temperature are the oxides' ability to conduct heat and the area of contact they make with the silicon. In particular, devices using $SiO_2$ underneath and $HfO_2$ above the channel show that the effective base width in contact with the back oxide largely governs how hot the channel becomes.

- Making the channel thicker relative to its width (increasing $T_{Si}/W_{Si}$) while keeping its cross-section identical causes higher channel temperatures, since the narrower base reduces heat-spreading. Thus, when tailoring channel shapes to alleviate self-heating, maximizing the channel's interface with its oxides is critical.

- The lateral doping pattern at the source/drain interfaces profoundly affects thermal behavior. Uniform doping leads to the hottest channels, whereas smoothly graded (analytical) profiles cut peak channel temperature by roughly 10%, boost Ion/Ioff nearly tenfold, and raise the threshold voltage by about 15% compared to abrupt, constant-doping designs.

## Acknowledgements

The authors express their gratitude to Dr. Jean Chamberlain Chedjou, Professor at the University of Klagenfurt (Austria), and Dr. Kumaravel Sivasankaran, Professor at Vellore Institute of Technology (India), for their collaboration in studying self-heating effects in nanoscale MOSFETs.

## Conflict of interest

The authors declare no conflict of interest.

## Author details

Ahmed Yusupov[1*], Atabek Atamuratov[2] and Makhkam Khalilloev[2]

1 Tashkent University of Information Technologies, Tashkent, Uzbekistan

2 Urgench State University, Urgench, Uzbekistan

*Address all correspondence to: a.yusupov@tuit.uz

## References

[1] Veeraraghavan S, Fossum JG. Shortchannel effects in SOIMOSFETs. IEEE Transactions on Electron Devices. 1989;**36**(3):522-528

[2] Ferain I, Colinge CA, Colinge JP. Multigate transistors as the future of classical metal-oxidesemiconductor field-effect transistor. Nature. 2011;**479**(7373):310-316

[3] Fossum JG, Zhou Z, Mathew L, B. Nguyen SOI versus bulk-silicon nanoscale FinFETs. Solid-State Electronics. 2010;**54**:86-89

[4] Park JT, Colinge JP, C.H. Diaz PiGate SOI MOSFET. IEEE Electron Device Letters. 2001;**22**(8):405-406

[5] Lee CW, Ferain I, Afzalian A, Yan R, Akhavan ND, Razavi P, et al. Performance estimation of junctionless multigate transistors. Solid-State Electronics. 2010;**54**:97-103

[6] Ahn W, Shin SH, Jiang C, Jiang H, Wahab MA, Alam MA. Integrated modeling of self-heating of confined geometry (FinFET, NWFET, and NSHFET) transistors and its implications for the reliability of sub-20 nm modern integrated circuits. Microelectronics Reliability. 2018;**81**:262-273

[7] Vasileska D, Goodnick SM, Raleva K. Proceedings of the 13th International Workshop on Computational Electronics, Beijing, China. Piscataway, NJ: IEEE; 2009. DOI: 10.1109/IWCE.2009.5091146

[8] Atamuratov AE, Saparov KS, Yusupov A, Chedjou JC. Combined influence of gate oxide and back oxide materials on self-heating and DIBL effect in 2D MOS2-based MOSFETs. Applied Sciences. 2023;**13**:6131. DOI: 10.3390/app13106131

[9] Atamuratov AE, Jabbarova BO, Khalilloev MM, Yusupov A, Loureriro AG. Self-heating effect in nanoscale SOI Junctionless FinFET with different geometries. In: 13th Spanish Conference on Electron Devices (CDE). Piscataway, NJ, USA: IEEE; 2021. DOI: 10.1109/CDE52135.2021.9455728

[10] Balasubbareddy M, Sivasankaran K, Atamuratov AE, Khalilloev MM. Optimization of vertically stacked nanosheet FET immune to self-heating. Micro and Nanostructures. 2023;**182**:207633. DOI: 10.1016/j.micrna.2023.207633, ISSN 2773-0123

[11] Atamuratov AE, Jabbarova BO, Khalilloev MM, Rajapov DR, Yusupov A, Chedjou JC, et al. Impact of source (drain) doping profiles and channel doping level on self-heating effect in FinFET. Micro and Nanostructures. 2025;**197**:208015. DOI: 10.1016/j.micrna.2024.208015, ISSN 2773-0123

[12] Atamuratov AE, Jabbarova BO, Khalilloev MM, Yusupov A. Sivasankaran. Influence of the gate oxide and Back oxide material types on self-heating effect in Junctionless FinFET. In: IEEE 25th International Conference of Young Professionals in Electron Devices and Materials (EDM). Piscataway, NJ, USA: IEEE; 2024. DOI: 10.1109/EDM61683.2024.10615156

[13] McDaid LJ, Hall S, Mellor PH, Eccleston W, J.C. Alderman physical origin of negative differential resistance in SOI transistors. Electron Letters. 1989;**25**(13):827-828

[14] Atamuratov AE, Jabbarova BO, Khalilloev MM, Yusupov A,

Sivasankaran K, Chedjou JC. Impact of the channel shape, back oxide and gate oxide layers on self-heating in nanoscale JL FINFET. Nanosystems: Physics, Chemistry, Mathematics. 2022;**13**(2):148-155

[15] Synopsys, Inc. Synopsys website. Sunnyvale, CA, USA: Synopsys; Available from: http://www. synopsys. com

[16] Loureiro A, Seoane N, Aldegunde MR, Valin R, Asenov A, Martinez A, et al. Implementation of the density gradient quantum corrections for 3-D simulations of multigate nanoscaled transistors. IEEE Transactions on Computer-Aided Design of Integrated Circuits and Systems. 2011;**30**(6):841-851

[17] Barraud S, Berthome M, Coquand R, Casse M, Ernst T, Samson M-P, et al. Scaling of Trigate Junctionless nanowire MOSFET with gate length down to 13 nm. IEEE Electron Device Letters. 2012;**33**(9):1225-1227

[18] Wang L, Brown AR, Nedjalkov M, Alexander C, Cheng B, Millar C, et al. Impact of self-heating on the statistical variability in bulk and SOI FinFETs. IEEE Transactions on Electron Devices. 2015;**62**(7):2106-2112. DOI: 10.1109/TED.2015.2436351

[19] Bury E, Kaczer B, Mitard J, Collaert N, Khatami NS, Aksamija Z, et al. Characterization of self-heating in high-mobility Ge FinFET pMOS devices. In: Proceedings of the 2015 Symposium on VLSI Technology (VLSI Technology), Kyoto, Japan. Piscataway, NJ, USA: IEEE; 2015. DOI: 10.1109/VLSIT.2015.7223703

[20] Atamuratov A, Saparov XS, Chedjou JC, Yusupov A, Kyamakya K. Self heating and DIBL effects in 2D MoS2 based MOSFET with different gate oxide and back oxide materials. In: 2022 International Conference on Information Science and Communications Technologies (ICISCT). Piscataway, NJ, USA: IEEE; 2022. DOI: 10.1109/ICISCT55600.2022.10146786

[21] Atamuratov AE, Jabbarova BO, Khalilloev MM, Yusupov A. The self-heating effect in Junctionless fin field-effect transistors based on silicon-on-insulator structures with Different Channel shapes. Technical Physics Letters. 2021;**47**(7):542-545

[22] Basker VS, Standaert T, Kawasaki H, Yeh CC, Maitra K, Yamashita T, et al. A 0.063 μm2 FinFET SRAM cell demonstration with conventional lithography using a novel integration scheme with aggressively scaled fin and gate pitch. In: Symposium on VLSI Technology Digest. Piscataway, NJ, USA: IEEE; 2010. pp. 19-20

[23] Bergman T, L, Lavine AS, Incropera FP. Fundamentals of Heat and Mass Transfer. 7th ed. Hoboken, NJ, USA: John Wiley & Sons, Incorporated; 2011. ISBN: 9781118137253

[24] Panzer M, A, Shandalov M, Rowlette JA, Oshima Y, Chen YW, McIntyre PC, et al. Thermal properties of ultrathin hafnium oxide gate dielectric films. IEEE Electron Device Letters. 2009;**30**(12):1269-1271. DOI: 10.1109/LED.2009.2032937

[25] Cahill DG, Allen TH. Thermal conductivity of sputtered and evaporated $SiO_2$ and $TiO_2$ optical coatings. Applied Physics Letters. 1994;**65**:309. DOI: 10.1063/1.112355

[26] Sze S, Ng KK. Physics of Semiconductor Devices: Third Edition. 3rd ed. Hoboken, NJ, USA: John Wiley & Sons, Inc.; 2006. DOI: 10.1002/9780470068328

Chapter 4

# Challenges and Benefits of Multi-Value Digital Circuit Design with GNRFET and CNTFET Technology

*Mahdieh Nayeri and Maryam Nayeri*

**Abstract**

The rapid advancement of digital electronics has been driven by the relentless scaling of transistors, as described by Moore's law. However, as silicon-based technologies approach, their physical limits, alternative materials, and circuit paradigms are being explored to sustain progress. This chapter investigates the potential of graphene nanoribbon field-effect transistors (GNRFETs) and carbon nanotube field-effect transistors (CNTFETs) as next-generation transistor technologies, particularly in the context of ternary logic circuits. Ternary logic offers advantages in information density, computational efficiency, and reduced interconnect complexity. GNRFETs leverage the tunable bandgap and high carrier mobility of graphene nanoribbons, while CNTFETs utilize the exceptional electrical properties of carbon nanotubes, including near-ballistic transport and high electron mobility. Key aspects such as energy efficiency, fabrication scalability, and device performance are analyzed, highlighting the strengths and challenges of each technology. Both technologies face significant challenges, including large-scale fabrication, device variability, and integration with existing silicon-based systems. The study concludes by exploring future prospects, such as hybrid architectures, scalable manufacturing techniques, and advanced applications in neuromorphic computing and optoelectronics. By addressing these challenges, GNRFETs and CNTFETs have the potential to revolutionize digital electronics, enabling faster, more energy-efficient, and compact devices for the next generation of computing systems.

**Keywords:** digital electronic, GNRFET, CNTFET, ternary logic circuits, next-generation transistor technologies

## 1. Introduction

Digital electronics have revolutionized information processing, enabling the proliferation of computers, communication devices, and embedded systems.

At the heart of this revolution lies the transistor, a semiconductor device that acts as the fundamental building block of digital circuits. Since its invention in 1947, the transistor has been the cornerstone of modern electronics, enabling dramatic improvements in speed, miniaturization, and energy efficiency. By functioning as a switch or amplifier, transistors allow for the precise control of electrical signals, making them indispensable in the design of logic gates, memory cells, and processors. The relentless scaling of transistors, as described by Moore's Law, has driven the exponential growth of computing power over the past several decades.

However, as conventional silicon-based transistors approach their physical limits, challenges such as power leakage, heat dissipation, and quantum effects have emerged, threatening further miniaturization and performance gains. This has spurred an intense search for next-generation transistor technologies that can overcome these limitations. Among the most promising candidates are carbon nanotube field-effect transistors (CNTFETs) and graphene nanoribbon field-effect transistors (GNRFETs). These novel devices leverage the exceptional electrical properties of carbon-based materials, such as high carrier mobility, excellent thermal conductivity, and atomic-scale scalability. CNTFETs, for instance, utilize carbon nanotubes as the channel material, offering superior performance compared to traditional silicon transistors, particularly in terms of speed and energy efficiency. Similarly, GNRFETs exploit graphene nanoribbons, which exhibit tunable bandgaps and high current-carrying capacity, making them ideal for high-performance applications.

Beyond advancements in materials, innovative circuit paradigms like ternary logic are being explored to further enhance computational power and reduce energy consumption. Ternary logic, which uses three states instead of the binary system's two, can potentially increase information density and reduce the number of transistors required for certain operations. This approach, combined with next-generation transistor technologies, could lead to a new era of computing that is faster, smaller, and more energy-efficient [1].

The continued evolution of transistor technology is critical to the future of digital electronics. Whether through the development of novel materials like carbon nanotubes and graphene or the exploration of alternative circuit architectures, transistors remain at the forefront of innovation. As researchers push the boundaries of what is possible, the transistor will continue to play a central role in shaping the next generation of devices, enabling advancements in artificial intelligence, quantum computing, and beyond. Together, these advances are paving the way for a future where digital electronics are not only faster and more compact but also more sustainable and capable of meeting the ever-growing demands of technology.

## 2. Foundations of digital electronics

Digital electronics form the cornerstone of modern computation and information processing by utilizing discrete voltage levels to represent logical states. This discrete approach contrasts with analog electronics, where signals vary continuously, enabling digital systems to achieve greater reliability and precision in signal interpretation and processing.

*DOI: http://dx.doi.org/10.5772/intechopen.1012236*

### 2.1 Binary and ternary logic

In binary logic, two distinct voltage levels represent the logical states 0 and 1. This binary system is the foundation of most digital systems due to its simplicity and robustness in noise immunity. The voltage thresholds for these states are carefully defined to ensure reliable signal interpretation, even in the presence of minor variations or interference [1].

Ternary logic introduces a third state, 2, which can increase data density and potentially improve the efficiency of certain computational tasks. For example, ternary systems can reduce the number of operations required for specific algorithms, leading to faster processing times [2]. However, ternary logic systems are more complex to design and implement compared to their binary counterparts, limiting their widespread adoption.

Basic Building Blocks: Logic Gates.

Digital systems are constructed from fundamental logic gates such as AND, OR, and NOT, which perform elementary operations on binary or ternary signals. These gates serve as the building blocks for more complex digital circuits. For instance:

The AND gate outputs a high signal only if all its inputs are high.

The OR gate outputs a high signal if at least one of its inputs is high.

The NOT gate inverts the input signal.

These gates are combined to create more sophisticated components like adders, multiplexers, and arithmetic logic units (ALUs), which enable advanced computational tasks such as arithmetic operations, data routing, and decision-making [2].

### 2.2 Memory elements and interconnects

Memory elements, such as flip-flops and latches, are essential for storing information in digital systems. These components maintain their state until explicitly changed, enabling sequential logic and data retention. Interconnects, on the other hand, facilitate communication between various circuit components, ensuring that data flows efficiently throughout the system [3].

### 2.3 Advancements beyond silicon

As traditional silicon-based technologies approach their physical limits, researchers are exploring alternative materials and device structures to drive the next generation of digital electronics. Notable innovations include:

Carbon Nanotube Field-Effect Transistors (CNTFETs): These transistors leverage the exceptional electrical properties of carbon nanotubes, offering higher speeds and lower power consumption compared to silicon-based transistors [4].

Graphene Nanoribbon Field-Effect Transistors (GNRFETs): Graphene nanoribbons provide a promising platform for ultra-scaled transistors due to their high electron mobility and atomic thickness, enabling further miniaturization and performance improvements [5].

These advancements are critical for addressing the growing demands for higher computational power, energy efficiency, and scalability in next-generation digital systems.

The foundations of digital electronics lie in the use of discrete voltage levels to represent logical states, with binary and ternary logic serving as the primary frameworks. Basic logic gates form the building blocks for complex digital systems, while memory elements and interconnects enable data storage and communication. Emerging technologies, such as CNTFETs and GNRFETs, are poised to overcome the limitations of traditional silicon, paving the way for continued innovation in digital electronics.

## 3. Multiple circuit design in digital systems

Multiple circuit design in digital systems focuses on integrating diverse logic functions and subsystems within a single chip or module, which is vital for complex, high-performance electronics. Combinational circuits—like adders and multiplexers—produce outputs based only on their current inputs, offering fast and reliable operations. Sequential circuits, including flip-flops and registers, rely on both present and past inputs, enabling memory and state retention. Integration techniques, such as hierarchical and modular design, as well as system-on-chip (SoC) methodologies, support greater scalability and design efficiency by organizing circuits into reusable and manageable blocks.

### 3.1 Challenges

Challenges in modern digital circuit design include managing power consumption, ensuring signal integrity, and overcoming scaling limitations in silicon-based devices. As device dimensions shrink, leakage currents and heat dissipation become major concerns. Maintaining reliable signal transmission and addressing short-channel effects are also critical as technology approaches fundamental material limits.

### 3.2 Combinational circuits: Speed and simplicity

Combinational circuits are fundamental building blocks in digital systems. These circuits, such as adders, multiplexers, and decoders, generate outputs based solely on their current inputs. Their key advantage lies in their speed and reliability, as they do not rely on past states or memory elements. For example, an adder circuit performs arithmetic operations instantly, making it indispensable in arithmetic logic units (ALUs) of processors. Similarly, multiplexers are widely used for data routing, enabling efficient selection and transmission of signals in complex systems.

### 3.3 Sequential circuits: Memory and state retention

In contrast, sequential circuits, such as flip-flops, registers, and counters, incorporate memory elements that allow them to depend on both current and past inputs. This capability is crucial for applications requiring state retention, timing control, and synchronization. For instance, flip-flops are the foundation of clocked systems, enabling the storage and transfer of binary data in microprocessors. Registers, which are collections of flip-flops, are used for temporary data storage, while counters are essential for tracking events or generating timing signals.

### 3.4 Integration techniques: Scalability and efficiency

To manage the complexity of modern digital systems, engineers employ advanced integration techniques that enhance scalability and design efficiency. Two prominent methodologies are:

#### *3.4.1 Hierarchical design*

This approach organizes circuits into layers of abstraction, allowing designers to focus on high-level functionality while delegating lower-level details to submodules. For example, a processor can be divided into functional units like the ALU, control unit, and memory interface, each of which is further broken down into smaller components.

#### *3.4.2 Modular design*

Modularity promotes the creation of reusable and interchangeable blocks, reducing redundancy and accelerating development. Standardized modules, such as memory blocks or communication interfaces, can be integrated across multiple designs, ensuring consistency and compatibility.

#### *3.4.3 System-on-Chip (SoC)*

SoC methodologies integrate entire systems—including processors, memory, I/O interfaces, and specialized accelerators—onto a single chip. This approach is particularly valuable in applications like smartphones and IoT devices, where compactness, power efficiency, and performance are critical.

## 4. Challenges in modern digital circuit design

Despite the advancements in integration techniques, designers face significant challenges as digital systems become more complex and device dimensions shrink:

### 4.1 Power consumption

As transistor densities increase, power consumption and heat dissipation become critical issues. Leakage currents, which occur even when devices are idle, contribute to energy waste and thermal challenges. Techniques such as dynamic voltage and frequency scaling (DVFS) and low-power design methodologies are employed to mitigate these effects.

### 4.2 Signal integrity

Ensuring reliable signal transmission is increasingly difficult as operating frequencies rise and interconnect lengths shrink. Signal degradation due to crosstalk, noise, and impedance mismatches can compromise system performance. Advanced shielding, routing strategies, and error correction codes are used to address these issues.

### 4.3 Scaling limitations

The relentless pursuit of Moore's law has pushed silicon-based devices to their physical limits. Short-channel effects, such as increased leakage and reduced gate control, undermine transistor performance at nanometer scales. Emerging technologies, including Fin FETs, gate-all-around (GAA) transistors, and alternative materials like graphene, are being explored to overcome these barriers.

### 4.4 Thermal management

High-density circuits generate significant heat, which can degrade performance and reliability. Effective thermal management strategies, such as heat sinks, thermal vias, and advanced cooling systems, are essential to maintain optimal operating conditions.

### 4.5 Design complexity

The growing intricacy of digital systems necessitates sophisticated design tools and methodologies. Electronic design automation (EDA) tools, machine learning algorithms, and hardware description languages (HDLs) like Verilog and VHDL are indispensable for managing complexity and ensuring design correctness.

### 4.6 Future directions

The future of multiple circuit design in digital systems lies in addressing these challenges while exploring new paradigms. Innovations such as 3D chip stacking, neuromorphic computing, and quantum computing promise to revolutionize the field. Additionally, the integration of AI and machine learning into design workflows is expected to enhance efficiency and enable the creation of more intelligent and adaptive systems.

In conclusion, multiple circuit design in digital systems is a dynamic and evolving discipline that plays a pivotal role in advancing technology. By leveraging combinational and sequential circuits, adopting scalable integration techniques, and tackling emerging challenges, engineers can continue to push the boundaries of what is possible in electronics.

## 5. Introduction to GNRFET technology

Graphene nanoribbon field-effect transistors (GNRFETs) represent a groundbreaking advancement in nanoelectronics, leveraging the unique properties of graphene to create highly efficient and versatile transistors. These devices utilize narrow strips of graphene, known as graphene nanoribbons (GNRs), as the channel material, offering significant advantages over conventional silicon-based transistors, such as metal-oxide-semiconductor field-effect transistors (MOSFETs). GNRFETs are poised to revolutionize the semiconductor industry by addressing key limitations of traditional technologies, including power consumption, speed, and scalability [5].

### 5.1 Key advantages of GNRFETs

#### *5.1.1 HighCarrierMobility*

Graphene is renowned for its exceptionally high electron mobility, which allows charge carriers to move through the material with minimal resistance. This property translates to faster switching speeds and improved performance in electronic devices.

#### *5.1.2 Tunable bandgap*

Unlike bulk graphene, which has a zero bandgap, graphene nanoribbons exhibit a bandgap that can be precisely tuned by altering their width. This tunability is critical for creating transistors with optimized on/off current ratios, making GNRFETs suitable for digital logic applications.

### *5.1.3 Low-power, high-speed operation*

The combination of high carrier mobility and a tunable bandgap enables GNRFETs to operate at lower power levels while maintaining high speeds. This makes them ideal for next-generation electronics, where energy efficiency and performance are paramount.

## 6. Structure of GNRFETs

The structure of GNRFETs is similar to that of traditional MOSFETs, consisting of three primary components:

- **Source**: The terminal through which charge carriers enter the channel.
- **Drain**: The terminal through which charge carriers exit the channel.
- **Gate**: The terminal that controls the flow of charge carriers by applying an electric field.

**Figure 1** shows graphene nanoribbon field-effect transistors. The key difference lies in the channel material: instead of silicon, GNRFETs use a graphene nanoribbon. This nanoribbon, which is only a few nanometers wide, serves as the conductive pathway between the source and drain. The gate electrode modulates the conductivity of the nanoribbon, enabling the transistor to switch between on and off states.

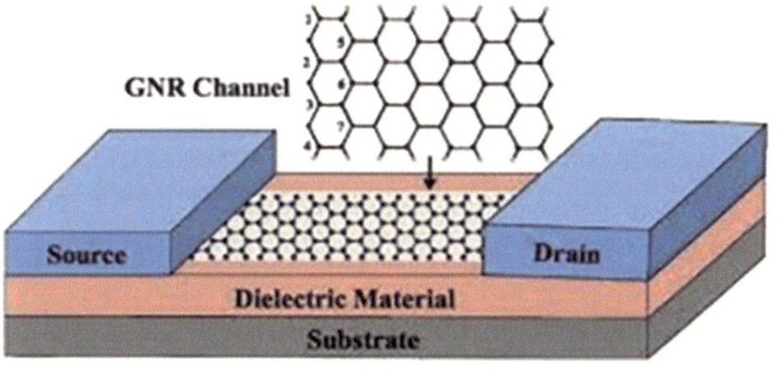


**Figure 1.**
*Graphene nanoribbon field-effect transistors [6].*

## 7. Relevance to ternary and multi-valued logic circuits

One of the most exciting applications of GNRFETs is their potential in multi-valued logic (MVL) circuits, particularly ternary logic, which operates with three distinct states (e.g., 0, 1, and 2) instead of the binary logic's two states (0 and 1). GNRFETs can be engineered to have multiple threshold voltages, allowing them to reliably distinguish between more than two states. This capability is achieved by carefully designing the graphene nanoribbon's width and edge structure, which influence the device's electronic properties.

Ternary circuits offer several advantages over binary systems, including higher information density, reduced interconnect complexity, and lower power consumption. GNRFETs' ability to support multi-valued logic positions them as a key enabler of future computing architectures that demand greater efficiency and scalability.

## 8. Broader implications and future prospects

The development of GNRFET technology has far-reaching implications for the semiconductor industry and beyond. By overcoming the limitations of silicon-based transistors, GNRFETs could enable the creation of ultra-compact, energy-efficient devices for applications ranging from high-performance computing to flexible and wearable electronics. Additionally, their compatibility with multi-valued logic opens new avenues for advancing artificial intelligence, quantum computing, and other emerging technologies.

However, challenges remain in the large-scale fabrication and integration of GNRFETs, including precise control over nanoribbon dimensions, edge defects, and device uniformity. Ongoing research and development efforts are focused on addressing these challenges, paving the way for the commercialization of GNRFET-based technologies.

In conclusion, graphene nanoribbon field-effect transistors represent a transformative step forward in nanoelectronics. With their unique properties and potential applications, GNRFETs are set to play a pivotal role in shaping the future of electronic devices and computing systems.

## 9. CNTFET technology overview

Carbon nanotube field-effect transistors (CNTFETs) are a cutting-edge technology that leverages carbon nanotubes (CNTs) as the channel material instead of traditional silicon. CNTFETs have garnered significant attention due to their exceptional electrical properties, which make them a promising candidate for next-generation electronic devices. Below is an expanded overview of CNTFET technology, including its structure, advantages, and relevance to ternary circuits.

### 9.1 Key advantages of CNTFETs

CNTFETs offer several advantages over conventional MOSFETs, including:

#### *9.1.1 High electron mobility*

Carbon nanotubes exhibit extremely high electron mobility, which translates to faster switching speeds and improved performance in high-frequency applications.

#### *9.1.2 Excellent on-off current ratio*

CNTFETs provide a superior on-off current ratio ($I_{ON}/I_{OFF}$), ensuring low leakage currents and efficient power consumption.

*DOI: http://dx.doi.org/10.5772/intechopen.1012236*

#### *9.1.3 Near-ballistic transport*

Electrons in CNTs experience minimal scattering, enabling near-ballistic transport. This results in reduced energy dissipation and enhanced device efficiency.

## 10. Structure of CNTFETs

The structure of CNTFETs is similar to that of traditional MOSFETs but with carbon nanotubes replacing the silicon channel. The key components include:

Source and Drain: Metallic contacts for electron injection and extraction.

Gate Electrode: Controls the flow of electrons through the channel.

Carbon Nanotube Channel: Single-walled nanotubes (SWNTs) or multi-walled nanotubes (MWNTs) serve as the conductive pathway. **Figure 2** illustrates the basic structure of a CNTFET.

### 10.1 Types of carbon nanotubes in CNTFETs

CNTFETs can utilize either single-walled nanotubes (SWNTs) or multi-walled nanotubes (MWNTs):

Single-Walled Nanotubes (SWNTs): Consist of a single layer of carbon atoms rolled into a tube. They offer superior electrical properties but are more challenging to fabricate.

Multi-Walled Nanotubes (MWNTs): Comprise multiple concentric layers of carbon nanotubes. They are easier to manufacture but may exhibit slightly reduced performance compared to SWNTs. **Figure 3** demonstrates single-walled and multi-walled CNTs.

#### *10.1.1 Relevance to ternary circuits*

CNTFETs are particularly well-suited for multi-valued logic (MVL) circuits, including ternary logic (three-state logic). The following factors contribute to their applicability:

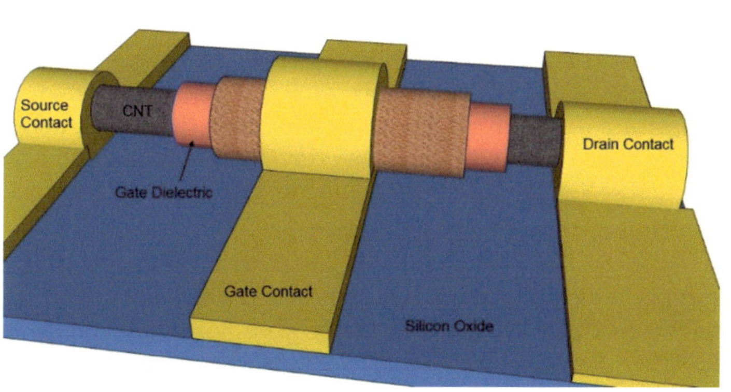


**Figure 2.**
*CNTFET structure [7].*

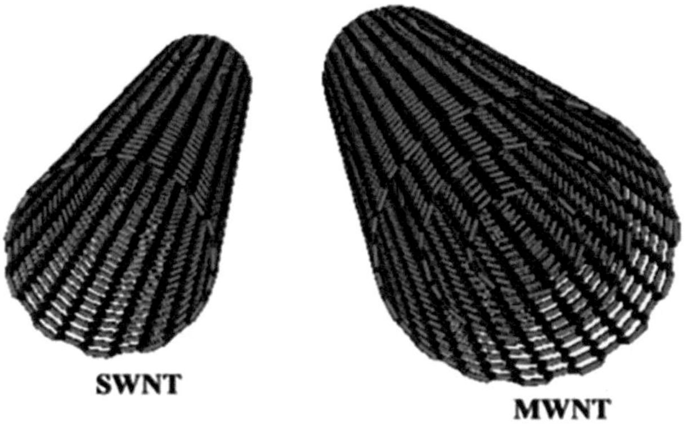


**Figure 3.**
*Single-walled and multi-walled CNTs [8].*

Threshold Voltage Control: The threshold voltage of CNTFETs can be precisely tuned by selecting the diameter and chirality of the carbon nanotubes. This enables the design of devices with multiple distinct threshold voltages, which is essential for MVL circuits.

Scalability: CNTFETs can be scaled down to extremely small dimensions while maintaining their electrical performance, making them ideal for high-density ternary logic circuits.

Energy Efficiency: The near-ballistic transport and low leakage currents of CNTFETs contribute to energy-efficient ternary circuit designs.

## 11. Ternary circuit design using GNRFET design principles

### 11.1 Multi-threshold GNRFETs for ternary logic

The key to implementing ternary logic with GNRFETs lies in the use of multi-threshold devices. By carefully engineering GNRFETs with distinct threshold voltages, designers can create three well-defined logic levels:

Low State (L): Represented by a low voltage level, corresponding to the “0” state.

Intermediate State (M): Represented by a medium voltage level, corresponding to the “1” state.

High State (H): Represented by a high voltage level, corresponding to the “2” state.

The design of multi-threshold GNRFETs involves precise control over the width and edge configuration of the graphene nanoribbons, as these parameters directly influence the bandgap and, consequently, the threshold voltage of the device.

### 11.2 Implementation of ternary logic gates

Ternary logic gates, such as ternary inverters, NAND, and NOR gates, can be implemented using GNRFETs by leveraging their multi-threshold characteristics. For example:

Ternary Inverter: A ternary inverter can be designed using three GNRFETs, each with a distinct threshold voltage. The inverter maps the input logic levels (L, M, H) to their complementary outputs (H, M, L), ensuring clear separation between the three states.

Ternary NAND and NOR Gates: These gates are constructed by combining multiple GNRFETs in specific configurations to perform ternary logic operations. The design focuses on optimizing the power consumption and area while maintaining robust state separation.

## 12. Performance factors in ternary GNRFET circuits

Several performance factors must be considered when designing ternary circuits with GNRFETs:

Switching Speed: The high carrier mobility of graphene enables fast switching, which is critical for high-speed ternary logic applications.

Static and Dynamic Power Dissipation: Efficient design techniques are required to minimize both static (leakage) and dynamic (switching) power consumption, ensuring energy-efficient operation.

Noise Margin: The ability to distinguish between the three logic levels in the presence of noise is crucial. Designers must ensure sufficient noise margins to maintain reliable operation.

Example: Ternary Inverter Design.

A practical example of ternary circuit design is the ternary inverter implemented using three GNRFETs:

Low-Threshold GNRFET: Responsible for transitioning the input from the low state (L) to the high state (H).

Medium-Threshold GNRFET: Handles the intermediate state (M), ensuring it maps correctly to the output.

High-Threshold GNRFET: Manages the transition from the high state (H) to the low state (L).

This design ensures clear state separation and efficient operation, demonstrating the potential of GNRFETs in ternary logic applications [9].

## 13. Ternary circuit design using CNTFET

Ternary circuit design using CNTFET design principles: Carbon nanotube field-effect transistors (CNTFETs) are emerging as a promising technology for ternary (three-valued) logic circuit design. Ternary logic, which operates on three distinct voltage levels (e.g., 0, 1, and 2), offers advantages over traditional binary logic in terms of reduced circuit complexity, improved data density, and enhanced computational efficiency. CNTFETs are particularly well-suited for ternary logic due to their unique electrical properties and the ability to precisely control their threshold voltage [10]. **Figure 4** demonstrates ternary circuits (not gate) based on CNTFETS.

Similarity to GNRFET, but with carbon nanotube properties CNTFET-based ternary circuit design shares some similarities with graphene nanoribbon field-effect transistors (GNRFETs), as both technologies leverage nanomaterials for improved performance. However, CNTFETs stand out due to the inherent properties of carbon nanotubes (CNTs), such as their high carrier mobility, excellent electrostatic control, and tunable bandgap. These properties enable precise threshold voltage control, which is critical for implementing ternary logic gates. By adjusting the diameter

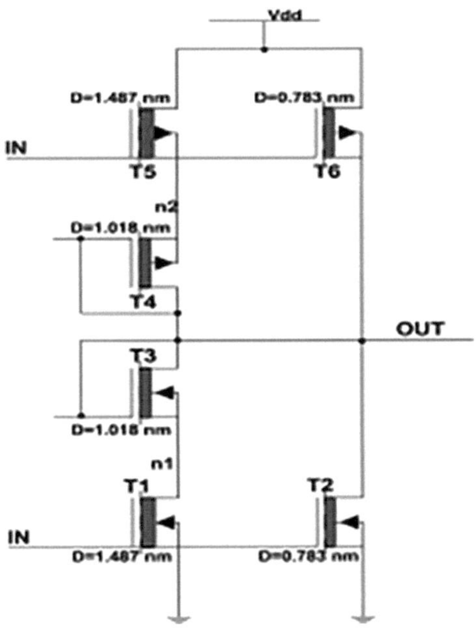


**Figure 4.**
*Ternary circuits based on CNTFETs [11].*

and chirality of the CNTs, designers can tailor the threshold voltage to differentiate between the three logic states effectively.

Established Models for CNTFET-Based Ternary Logic Circuits Over the years, researchers have developed well-established models and design methodologies for CNTFET-based ternary logic circuits. These include:

Decoders: Efficient ternary decoders can be designed using CNTFETs to reduce the number of transistors required compared to binary decoders, leading to area and power savings.

Adders: Ternary adders based on CNTFETs demonstrate improved performance in terms of speed and energy efficiency, making them suitable for arithmetic operations in advanced computing systems.

Memory Cells: CNTFET-based ternary memory cells offer higher storage density and reduced leakage currents, enabling more compact and energy-efficient memory architectures.

Performance Factors in CNTFET-Based Ternary Circuit Design: When designing ternary circuits using CNTFETs, several key performance factors must be considered to ensure optimal functionality and scalability:

Energy Efficiency: CNTFETs are inherently energy-efficient due to their low leakage currents and high on/off current ratios. Ternary logic further enhances energy efficiency by reducing the number of logic gates and interconnects required for complex computations, leading to lower dynamic and static power consumption.

Fabrication Scalability: While CNTFET technology is still evolving, significant progress has been made in developing scalable fabrication techniques. Challenges such as precise alignment of carbon nanotubes and uniform growth remain, but advancements in lithography and self-assembly methods are addressing these issues. Scalable fabrication is crucial for the widespread adoption of CNTFET-based ternary circuits in industrial applications.

## 14. Tolerance to process variations

Process variations during fabrication can impact the performance of CNTFETs. However, the tunable properties of CNTs provide a degree of resilience to these variations. Robust design techniques, such as redundancy and adaptive threshold control, can further enhance the tolerance of ternary circuits to process-induced uncertainties [12].

## 15. Comparative analysis: GNRFET vs. CNTFET in ternary logic applications

### 15.1 Overview

Graphene nanoribbon field-effect transistors (GNRFETs) and carbon nanotube field-effect transistors (CNTFETs) are emerging nanoscale technologies that hold significant promise for ternary logic applications. Ternary logic, which operates with three distinct states (0, 1, and 2), offers advantages over traditional binary logic in terms of information density and computational efficiency. Both GNRFETs and CNTFETs are being explored as potential successors to CMOS technology, particularly for low-power, high-speed applications. This analysis compares the key features of these two technologies, highlighting their strengths, challenges, and potential for ternary logic implementation.

### 15.2 Feature comparison

See **Table 1** describes the feature comparison between CNTFET and GNRFET.

| Feature | GNRFET | CNTFET |
|---|---|---|
| Carrier Mobility | High | Very high |
| Bandgap Tunability | Excellent (via width control) | Good (via diameter/chirality control) |
| Fabrication Challenges | Edge roughness, uniformity issues | Alignment difficulties, purity concerns |
| Device Scalability | Promising | Advanced but challenging |
| Power Consumption | Low (theoretical) | Very low |
| Ternary Logic Potential | High (multi-threshold operation) | High (multi-threshold operation) |

**Table 1.**
*Feature comparison between CNTFET and GNRFET.*

### 15.3 Key points

#### *15.3.1 Energy efficiency and speed*

Both GNRFETs and CNTFETs offer significant improvements in energy efficiency and speed compared to traditional CMOS technology. This makes them attractive candidates for next-generation computing, particularly in applications where power consumption and performance are critical.

1. Bandgap Tunability
2. GNRFETs: Excel in bandgap tunability, as the bandgap can be precisely controlled by adjusting the width of the graphene nanoribbon. This feature is particularly advantageous for designing multi-threshold devices required for ternary logic.
3. CNTFETs: While bandgap tunability is achievable through control of the nanotube's diameter and chirality, it is less precise compared to GNRFETs.
4. Carrier Mobility
5. CNTFETs: Exhibit very high carrier mobility, which translates to faster switching speeds and better performance in high-frequency applications.
6. GNRFETs: Also offer high carrier mobility, though slightly lower than CNTFETs.
7. Fabrication Challenges
8. GNRFETs: Face challenges related to edge roughness and achieving uniformity in nanoribbon width, which can impact device performance and yield.
9. CNTFETs: Struggle with alignment and purity issues during fabrication, as achieving precise placement and eliminating metallic nanotubes remain significant hurdles.
10. Device Scalability
11. GNRFETs: Show promise in scalability due to their planar structure, which is compatible with existing semiconductor fabrication techniques.
12. CNTFETs: While advanced in terms of scalability, challenges in achieving precise alignment and density control limit their current practicality.
13. Power Consumption
14. Both technologies are inherently low power, with CNTFETs having a slight edge due to their superior carrier mobility and electrostatic control.

## 16. Ternary logic potential

Both GNRFETs and CNTFETs are highly suitable for ternary logic applications due to their ability to support multi-threshold operation. This enables the implementation of three distinct logic states, making them ideal for future computing paradigms.

## 17. Challenges and future prospects

Large-scale, defect-free fabrication: One of the most significant hurdles in the development of graphene nanoribbons (GNRs) and carbon nanotubes (CNTs) is achieving large-scale, defect-free production.

GNRs: The fabrication of GNRs often results in imperfections such as edge roughness, which can introduce scattering centers and degrade electrical performance. Achieving precise control over the width and edge structure of GNRs is critical for maintaining their electronic properties.

CNTs: CNTs face challenges such as misalignment during growth and the presence of metallic impurities. Metallic CNTs, as opposed to semiconducting ones, can short-circuit devices, while misalignment can lead to inconsistent electrical contacts. These issues reduce device yield and performance.

Device-to-device variability: Variations in the size, shape, or chirality of GNRs and CNTs during fabrication lead to inconsistent electrical properties.

GNRFETs: Variations in ribbon width and edge structure can cause shifts in threshold voltage, making it difficult to achieve uniform device behavior.

CNTFETs: Chirality variations in CNTs result in differences in bandgap and carrier mobility, leading to unpredictable device performance. This variability complicates the design of reliable circuits and systems.

Integration with silicon: Integrating GNRFETs and CNTFETs into existing silicon CMOS technology is a complex challenge.

Thermal budgets: The high temperatures required for synthesizing GNRs and CNTs may exceed the thermal limits of silicon substrates, potentially damaging existing CMOS layers.

Interconnects and packaging: Developing compatible interconnects and packaging methods is essential for seamless integration. This includes ensuring that the new materials can interface with traditional metal interconnects and meet the mechanical and thermal requirements of modern packaging techniques [13].

### 17.1 Future prospects

Hybrid architectures: Combining CNTFETs or GNRFETs with traditional silicon devices offers a promising path forward.

High-speed logic and memory: Hybrid circuits could leverage the high carrier mobility of CNTFETs and the tunable bandgap of GNRFETs for high-speed logic operations, while relying on mature silicon technology for memory systems.

Multi-functional devices: Such architectures could enable new functionalities, such as reconfigurable circuits or systems-on-chip (SoCs) that integrate processing, memory, and sensing capabilities.

Scalable manufacturing: Advances in synthesis techniques are crucial for the mass production of GNRs and CNTs.

Chemical vapor deposition (CVD): Improvements in CVD methods could enable the growth of uniform, defect-free nanomaterials at wafer scale, making large-scale fabrication feasible.

Self-assembly and directed growth: Techniques that promote self-assembly or directed growth of GNRs and CNTs could reduce variability and improve alignment, enhancing device uniformity.

Advanced applications: As fabrication techniques mature, GNRFETs and CNTFETs could enable a range of advanced applications.

Ternary logic circuits: These devices are well-suited for ternary logic, which offers higher information density and energy efficiency compared to binary logic. This could lead to ultra-dense, low-power processors, and memory arrays.

Neuromorphic computing: The unique properties of GNRs and CNTs, such as their high speed and low power consumption, make them ideal candidates for neuromorphic systems that mimic the brain's architecture for efficient and adaptive computing.

Optoelectronics and sensors: The tunable bandgap of GNRs and the exceptional electrical properties of CNTs could be harnessed for next-generation optoelectronic devices and highly sensitive sensors [14].

## 18. Conclusion

GNRFET and CNTFET technologies represent groundbreaking advancements in the field of digital electronics, offering the potential for higher performance and greater energy efficiency compared to traditional silicon CMOS.

GNRFETs: Their ability to tune the bandgap through width control makes them versatile for applications in nanoscale transistors, sensors, and optoelectronic devices.

CNTFETs: With their exceptional carrier mobility and ultra-low power consumption, CNTFETs are ideal for high-speed logic and low-power circuits.

However, significant challenges remain, particularly in the areas of large-scale fabrication, device uniformity, and integration with existing silicon technology. Overcoming these challenges will require innovations in materials science, manufacturing processes, and circuit design.

Interdisciplinary research: Collaboration across disciplines will be essential to address the multifaceted nature of these challenges.

Hybrid approaches: Combining GNRFETs and CNTFETs with silicon CMOS in hybrid architectures may accelerate their adoption and enable new functionalities in electronic systems.

As these technologies mature, they have the potential to revolutionize digital electronics, paving the way for more compact, efficient, and powerful devices in applications ranging from high-speed computing to neuromorphic systems and beyond.

## Author details

Mahdieh Nayeri[1*] and Maryam Nayeri[2]

1 Department of Computer Engineering, Institute of Higher Education Pooyandegan Danesh, Mazandaran, Iran

2 Department of Electrical Engineering, Ya.C., Islamic Azad University, Yazd, Iran

*Address all correspondence to: nayeri2020@gmail.com

## References

[1] Hurst. Multiple-valued logic—Its status and its future. IEEE Transactions on Computers. 1984;**100**(12):1160-1179

[2] Yang L, Park CH, Son YW, Cohen ML, Louie SG. Quasiparticle energies and band gaps in graphene nanoribbons. Physical Review Letters. 2007;**99**(18):186801

[3] Geim AK, Novoselov KS. The rise of graphene. Nature Materials. 2007;**6**(3):183-191

[4] Schwierz F. Graphene transistors. Nature Nanotechnology. 2010;**5**(7):487-496

[5] Lin S-Y, Thanh Thuy Tran N, Chang S-L, Su W-P, Lin M-F. Structure- and Adatom-Enriched Essential Properties of Graphene Nanoribbons. 1st ed. CRC Press; 2018. DOI: 10.1201/9780429400650

[6] Chen Y-Y, Sangai A, Rogachev A, Gholipour M, Iannaccone G, Fiori G, et al. A SPICE-compatible model of MOS-type graphene nano-ribbon field-effect transistors enabling gate-and circuit-level delay and power analysis under process variation. IEEE Transactions on Nanotechnology. 2015;**14**(6):1068-1082

[7] Rahmati S, Farshidi E, Ganji J. Low energy and area efficient quaternary multiplier with carbon nanotube field effect transistors. ETRI Journal. 2021;**43**(4):717-727

[8] Krishna R, Hari MN, Chandraprabha KS, Krishna Murthy TP, Manjunatha C, Girish Kumar S. Carbon nanotubes and graphene-based materials for adsorptive removal of metal ions–A review on surface functionalization and related adsorption mechanism. Applied Surface Science Advances. 2023;**16**:100431

[9] Basha SJ, Elbarbary A, Mahmoud HA, Venkatramana P. High performance and PVT variation tolerant design for a ternary multiplier using GNRFET technology. Circuits, Systems, and Signal Processing. 11 Mar 2025:1-27

[10] Moaiyeri MH, Doostaregan A, Navi K. Design of energy-efficient and robust ternary circuits for nanotechnology. IET Circuits, Devices & Systems. 2011;**5**(4):285-296

[11] Lin S, Kim Y-B, Lombardi F. CNTFET-based design of ternary logic gates and arithmetic circuits. IEEE Transactions on Nanotechnology. 2009;**10**(2):217-225

[12] Franklin AD, Luisier M, Han S-J, Tulevski G, Breslin CM, Gignac L, et al. Sub-10 nm carbon nanotube transistor. Nano Letters. 2012;**12**(2):758-762

[13] Abbasian E, Aminzadeh A, Anvari ST. GNRFET-and CNTFET-based designs of highly efficient 22 T unbalanced single-trit ternary multiplier cell. Arabian Journal for Science and Engineering. 2023;**48**(11):15337-15352

[14] Niu W, Sopp S, Lodi A, Gee A, Kong F, Pei T, et al. Exceptionally clean single-electron transistors from solutions of molecular graphene nanoribbons. Nature Materials. 2023;**22**(2):180-185

Chapter 5

# Performance Assessment of GaSb-Si Vertical TFET with Triple Metal Gate for Low Power Application

*Debasish Mohanta, Sruti Suvadarsini Singh and Prasanna Kumar Sahu*


## Abstract

This chapter represents the simulation study of a GaSb-Si triple metal gate vertical tunnel field-effect transistor (TMG-VTFET). The TMG-VTFET has distinct work function which is analysed using technology CAD (TCAD) in this chapter. In this proposed device structure, the metal gate is segregated into three regions. In this study, we have obtained better switching characteristics of TMG-VTFET by using a narrow bandgap material in the source region and improved electrostatic control on the channel because of triple metal gate. It exhibits an $I_{on}$ of $1.75 \times 10^{-4} A/\mu m$, an $I_{off}$ of $8.835 \times 10^{-19} A/\mu m$, an $I_{on}/I_{off}$ ratio of $1.9 \times 10^{14}$, and subthreshold slope (SS) of 22.1 $mV/dec$ at $V_{DS} = 0.5V$. The heterojunction gate dielectric assists in upgrading the properties (ON current and OFF current) of the proposed device. There are three distinct metal work functions used in the channel junction that generates potential difference in the channel junction, thereby decreasing the current flowing through the drain. In this work, ambipolarity is being controlled by the structure of triple metal gate. We have also explored the influence of metal work functions and gate oxide on the proposed device characteristics. The simulation results demonstrate that the proposed TMG-vertical TFET exhibits low power consumption and fast switching characteristics, making it a promising candidate for digital applications.


**Keywords:** GaSb-Si, TMG, VTFET, TFET, SS, CG, AG, TG

## 1. Introduction

The physical dimensions of integrated circuit transistors have been successfully reduced in accordance with Moore's law. However, the aggressive miniaturisation of transistors is hampered by a few unavoidable issues, such as the short-channel effects and the steadily rising power consumption. Tunnel field effect transistors (TFETs), popularly known as green transistor, employ a band to band tunnelling (BTBT) phenomena rather than the thermionic-emission process that have been studied as most promising devices in low-power applications in order to address the technical issues [1–3].

TFETs have recently come into focus because of their exceptional subthreshold characteristics and low-power operating capabilities [4]. Unlike MOSFET, where current is created by thermionic emission, TFETs can achieve a steep subthreshold slope (SS) owing to band-to-band tunneling (BTBT) of carriers from the source to the channel [5]. The SS of the TFET, sometimes referred to as a "green transistor," is significantly lower at room temperature than the fundamental physical constraint on the subthreshold swing (SS) of 60 mV/decade for traditional MOSFETs [6–8]. Because of their high tunnelling barrier width, which stops electrons from tunnelling at the source-channel junction with decreasing gate bias, TFETs have incredibly low leakage currents.

Low ON-state drive current($I_{ON}$), inadequate subthreshold slope, and high ambipolar current are some of the TFET's main drawbacks [9, 10]. A source region composed of innovative materials, such as SiGe, *Ge*, or group III–V low bandgap materials [11, 12], with a gate-all-around structure [13–16], was employed to overcome these technological obstacles in order to produce a steeper subthreshold slope (SS) and a higher $I_{ON}$.

Researchers have proposed a number of methods to improve the ON-state current, for example, with the help of using low band gap material in the source region [17–19], high-K dielectric materials [5, 20], highly doped pockets region close proximity with the source [21, 22], hetero gate dielectric structures [23, 24] etc.

In this work, GaSb-Si triple metal gate VTFET with distinct work function is proposed. This work uses a triple metal gate construction to control ambipolarity. The chapter's first portion discusses recent TFETs. The device structure section highlights the device portion. The GaSb-Si triple metal gate TFET devices are covered in the results and discussion section. Finally, the conclusion section brings the chapter to a close.

## 2. Device structure

**Figure 1** illustrates the cross-sectional sight of GaSb-Si heterojunction triple metal gate vertical TFET. There are three metal gate sections which have separate work functions. Therefore, the channel length of the device evaluated as $L_{Ch} = L_1 + L_2 + L_3$ is totally 20 nm. The parameters of the triple metal gate vertical TFET are thickness of drain, source, channel, oxide layers, three gate regions, and doping concentrations which are used in simulation listed in **Table 1**.

The material which is utilised in the source side is GaSb. The matter utilised in drain and channel region is silicon. The upper region which is the source is being doped heavily. The channel region which is placed inside of gate and oxide layer is lightly doped. The TFET structure, which is proposed, composed of a silicon substrate having thickness of 10 nm. Source area is added with $P^+$ impurity with a concentration of $10^{20}/cm^3$. The channel region is doped with $5 \times 10^{16}/cm^3$ impurity concentration. The length ($L_S$) of the source side is 20 nm. The length ($L_{Ch}$) of the channel region and drain ($L_D$) are 20 nm and 40 nm, respectively. Drain region of the device is uniformly doped with $N^+$ concentration of $5 \times 10^{18}/cm^3$. Further, a 2 nm thickness of high permittivity $HfO_2$ layer is used as channel. The gate of the proposed device is partitioned into three segments which are named as tunnel gate (TG), control gate (CG), and auxiliary gate (AG), and lengths of TG, CG, and AG are 5 nm, 10 nm, and 5 nm, respectively. Three separate metals with distinct work functions are being used in this proposed structure. Using separate metals thereby enhances the on-state current $I_{on}$. Near the source, the TG regulates the source and energy bands (EB) in the channel and helps the electrons for passage between the source and channel region; therefore, there is an improvement in $I_{on}$. Near the drain, the AG's function is to

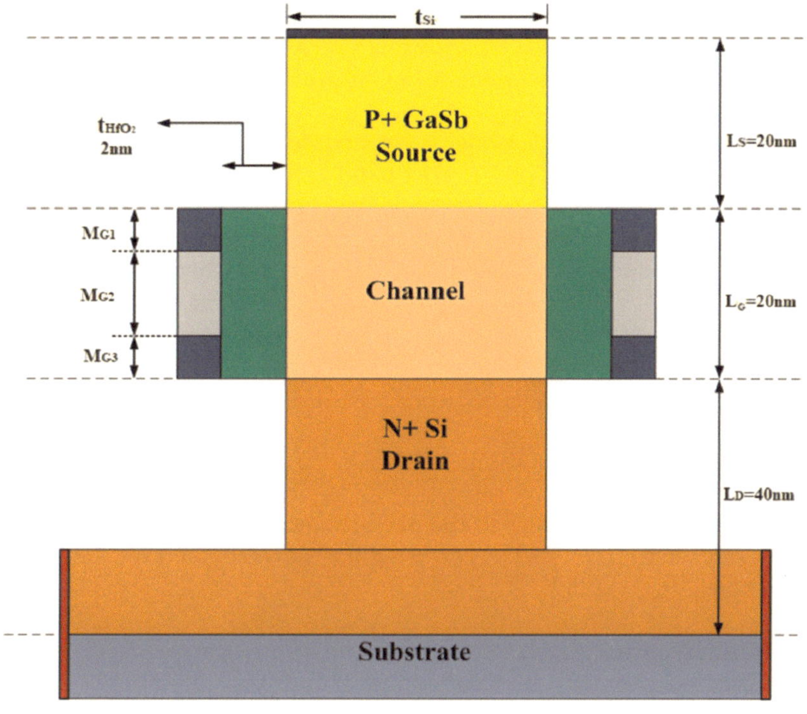


**Figure 1.**
*Schematic cross sectional view of GaAs-Si-TMG-VTFET device.*

| Parameters | Symbol | Value |
|---|---|---|
| Source doping concentration | $N_S$ | $1 \times 10^{16}/cm^3$ |
| Channel doping concentration | $N_{Ch}$ | $5 \times 10^{16}/cm^3$ |
| Drain doping concentration | $N_D$ | $5 \times 10^{18}/cm^3$ |
| Channel length | $L_{Ch}$ | 20 nm |
| Source length | $L_S$ | 20 nm |
| Drain length | $L_D$ | 40 nm |
| Gate oxide thickness | $T_{HfO_2}$ | 2 nm |
| TG length | $L_{TG}$ | 5 nm |
| CG length | $L_{CG}$ | 10 nm |
| AG length | $L_{AG}$ | 5 nm |
| TG work function | WF1 | 4.3 eV |
| CG work function | WF2 | 4.5 eV |
| AG work function | WF3 | 4.7 eV |

**Table 1.**
*Device parameters which are used in the process of simulation.*

maintain the minimum $I_{off}$ current. CG is the metal gate which is between the AG and the TG. The work function of TG's is smaller than the other gates in order to attain improved $I_{on}$, better SS, and reduced $I_{off}$ [25].

The TG by enhancing carrier tunnelling, improves ON-state current, whereas, AG near drain side decremented the OFF current. The AG work function is maintained more than TG to upgrade $I_{on}$, decrement $I_{off}$, improve SS, and reduce ambipolar current.

The dynamic non local BTBT model is essential to model the device using silvaco TCAD simulation. For this purpose, energy band gradient is being assessed. There is a non-uniform electric field in hetero-junction structure. The Shockley-Read-Hall model (SRH) is essential for simulation, generation, and also recombination at the semiconductor–semiconductor region and semiconductor–insulator region. Trap-assisted model (TAM) is also utilised to characterise the traps at the interface. Fermi-Dirac statistics, bandgap narrowing (BGN), field-dependent mobility models, and drift diffusion model are utilised for TCAD simulations.

## 3. Results and discussion

In this section, we have displayed the performance of TMG-VTFET including GaSb as the source material. The DC performance is investigated for the TFET at drain to source voltage ($V_{DS}$) is 0.5 V.

### 3.1 Energy band diagram

The energy band figure of the recommended structure in OFF and ON region is shown in **Figure 2**. When $V_{GS} = 0V$, the valence band (VB) of the source region remains above the conduction band (CB) of the channel region. Therefore, BTBT is prohibited. Gradually, incrementing $V_{GS}$ modulates the charge density, thereby driving the conduction band in the channel. As depicted in **Figure 2**, the CB of channel lines up with that of the VB of the source, when $V_{GS}$ increases to a high voltage level. Due to this, band bending takes place at the source region. Therefore, tunnelling of carriers occurs at the VB of the source region to the CB of the channel. The carriers

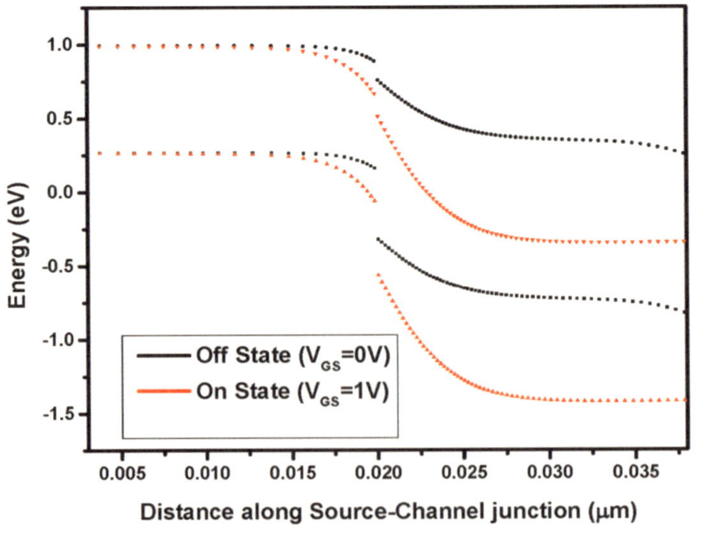


**Figure 2.**
*Energy band structure of the proposed structure.*

which are tunnelled are collected at the drain region by supplying drain voltage. The built-in potential barrier is generated due to the distinct work functions of gate metal. $V_{GS}$ can effectively regulate the pinnacle of the channel barrier. When the device is not conducting, the built-in potential barriers that corresponds to the work function WF1 and WF2 broaden the tunnelling barrier, and hence the likelihood of electrons tunnelling is limited. In the ON-state, with the applied gate voltage, the potential barrier is apparently lowered. Thereafter, the tunnelling width influentially remains low, allowing the carriers to tunnel with proficiency. The band is bending downwards which is attributable near the source region which is having a smaller work function WF1 consequence in a considerable electric field neighbouring the tunnelling source-channel region; therefore, the width of the tunnel shortens from the source to channel.

### 3.2 BTBT

**Figure 3** shows the rate of BTBT the VTFET device. It is obvious that the GaSb-Si TMG-VTFET device delivers very high BTBT rate. This is because of the very high value of electric field at the tunnelling junction.

The nonlocal BTBT is depicted in **Figure 4**. It has been noticed that, in on state, the nonlocal BTBT rate in the channel and source region increases.

### 3.3 Electric field and potential

**Figure 5** shows electrical field affected when $V_{GS}$ is applied. As $V_{GS}$ is increased from 0 V to 1 V, the enhanced BTBT results and the electrical field in the source-channel region grow as $V_{GS}$ increases. Subsequently, near the drain and channel junction, the electrical field is suppressed, and therefore, an insignificant flow of ambipolar current results.

**Figure 6** demonstrates the surface potential of the proposed device. At the surface, the inversion layer results for an incremented value of surface potential. The surface potential near the source region is decreased, thereby disallowing the occurrence of negative inversion layer. The below figure depicts that the surface potential close to the source junction is low and thereafter increases. The surface potential tunnelling junction is being impacted by the potential applied at the drain-source.

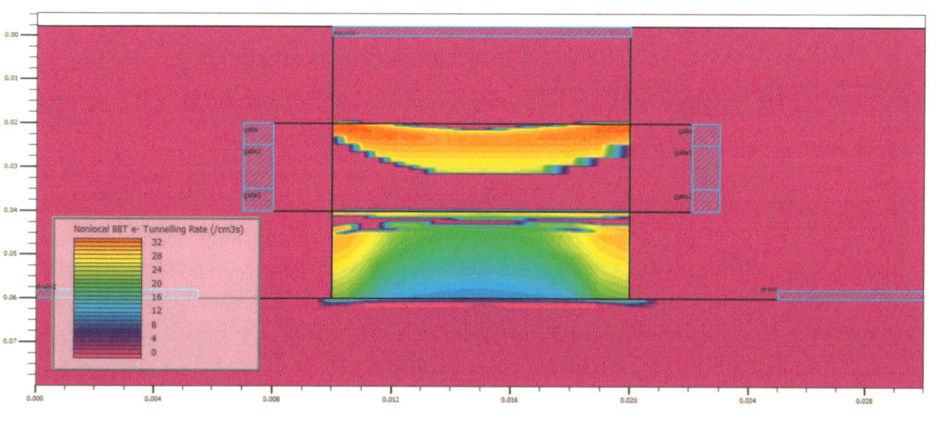


**Figure 3.**
*On state contour plot of BTBT rate.*

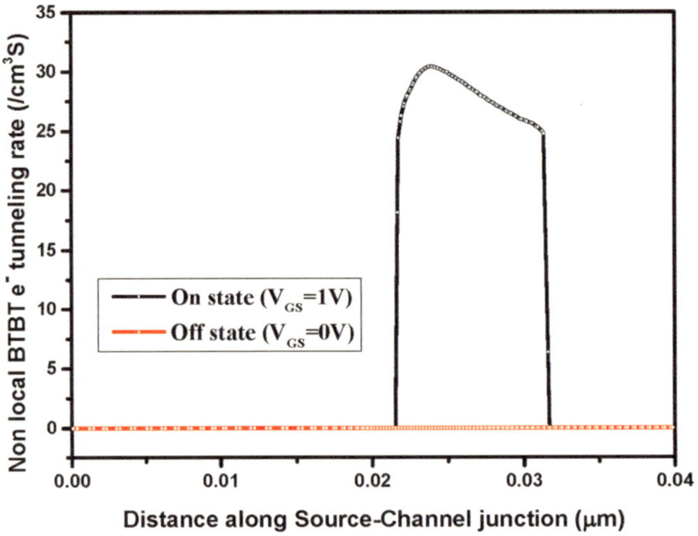


**Figure 4.**
*Nonlocal band to band tunnelling rate (BTBT) of the device.*

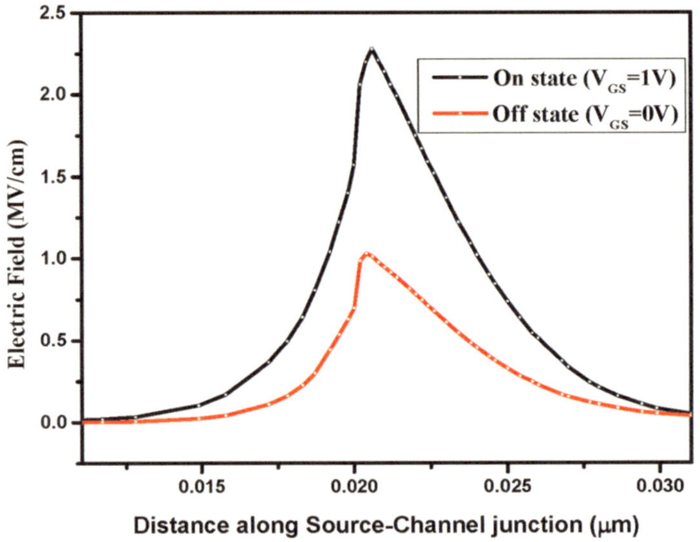


**Figure 5.**
*Electric field versus position of the proposed TMG-TFET device.*

## 3.4 DC performance analysis

The model parameters are calibrated against the experimental data reported in [25] for authenticity of the simulation work in **Figure 7**. The ideal values of three different work functions that are used at gate are WF1 = 4.3 eV, WF2 = 4.5 eV and WF3 = 4.7 eV. The value of on state current of this device $I_{on}$is $1.75 \times 10^{-4} A/_{\mu m}$ and average subthreshold slope at 25°C is 22.1 mV/decade at $V_{DS}$ = 0.5 V. The average subthreshold slope is paramount in performing switching characteristics of the proposed device. It is being measured using the gate to source voltage ($V_{GS}$). It is

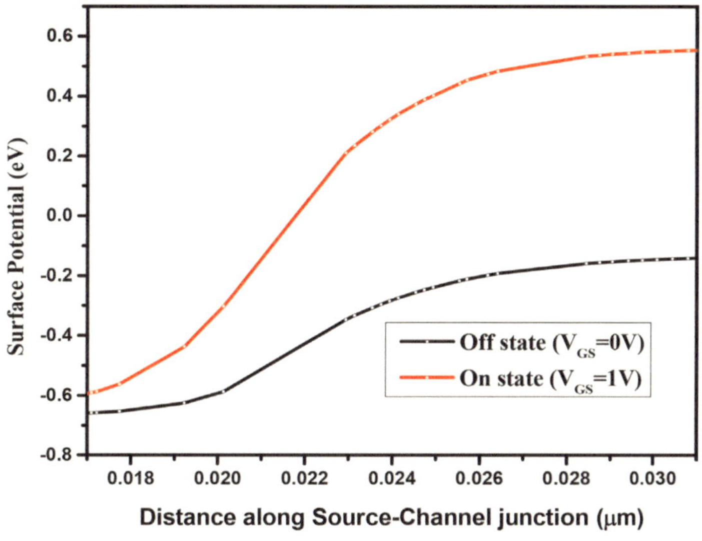


**Figure 6.**
*Surface potential for the proposed device.*

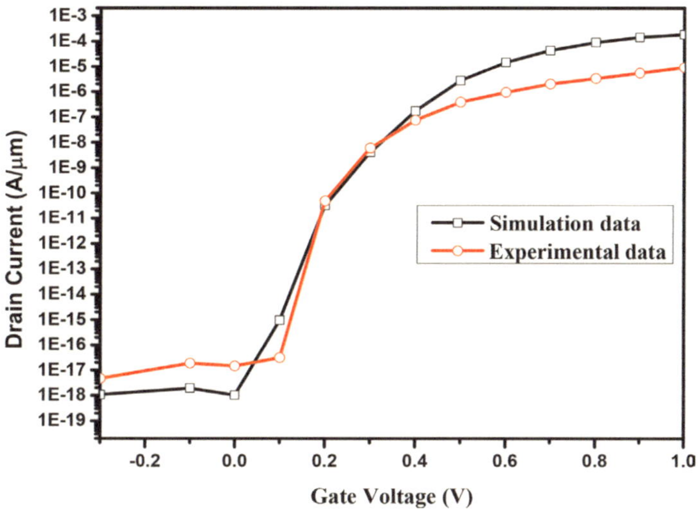


**Figure 7.**
*Calibration of the simulated data against the experimental results reported in [25].*

considered between 0 V and threshold voltage ($V_{TH}$). At $V_{GS}$ = 0 V, the drain current $I_{DS}$ starts to improve. The turn on voltage of the device is 0.38 V. When the applied gate voltage increases, the threshold voltage also begins to increase, thereby the subthreshold slope increases. It is preferable to have improved on state current with minimum gate voltage for good switching action of the device.

### *3.4.1 Variation of work function in different metals*

**Figure 8** shows the $V_{GS} - I_{DS}$ curves of GaSb-Si-TMG-TFET for different values of WF1 and the values of WF2 and WF3 are constant with 4.5 eV and 4.7 eV,

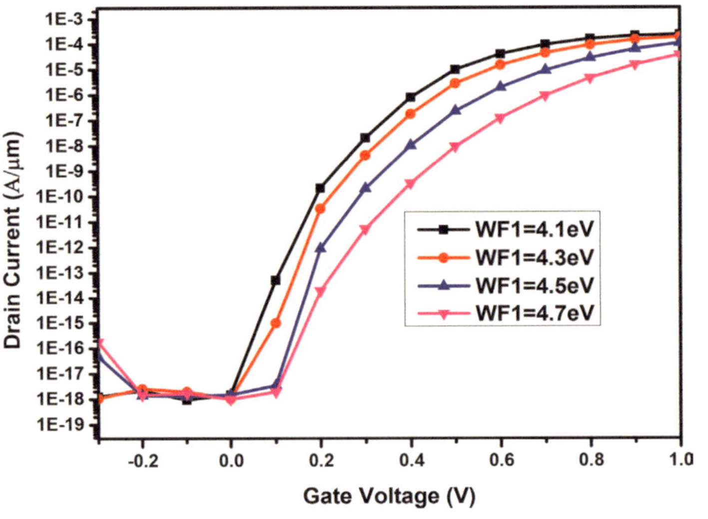


**Figure 8.**
*$V_{GS} - I_{DS}$ curve of GaSb-Si-TMG-TFET for different values of WF1 with WF2 = 4.5 eV and WF3 = 4.7 eV.*

| Work function (eV) | $V_{TH}$ | $I_{on}$ (A/μm) | $I_{off}$ (A/μm) | $I_{on}/I_{off}$ | SS(mV/decade) |
|---|---|---|---|---|---|
| 4.1 | 0.345 | $2.21 \times 10^{-4}$ | $1.025 \times 10^{-18}$ | $2.16 \times 10^{14}$ | 21.44 |
| 4.3 | 0.386 | $1.75 \times 10^{-4}$ | $8.835 \times 10^{-19}$ | $1.986 \times 10^{14}$ | 22.1 |
| 4.5 | 0.475 | $1.01 \times 10^{-4}$ | $1.025 \times 10^{-18}$ | $9.882 \times 10^{13}$ | 18 |
| 4.7 | 0.594 | $3.48 \times 10^{-5}$ | $1.57 \times 10^{-18}$ | $2.217 \times 10^{13}$ | 24.27 |

**Table 2.**
*Influence of changing WF1 on the switching characteristics of GaSb-Si-TMG-TFET with WF2 = 4.5 eV and WF3 = 4.7 eV.*

respectively. The voltage applied $V_{DS}$ is 0.5 V. From the figure, it is noticed that when WF1 increases, on state current of the TFET device decreases. As WF1 increases, the probability of tunnelling of carriers decreases. It is superior to have low value of WF1 for improved on state current. **Table 2** depicts the effects of varying WF1 with a constant value of WF2 and WF3.

The gate electrode work function on the source side plays a critical role in TFET operation for digital applications due to its close proximity to the tunneling region of the device. As illustrated in **Table 2**, it can significantly influence the value of on current, off current, and the subthreshold slope steepness. In **Figure 4**, the I-V curve of GaAs-Si-TMG-TFET with numerous work functions value of source side gate is demonstrated. A low metal work function is employed near the source region. Due to the work function difference across the gate, an electric field is generated in the channel, thereby increasing BTBT. Therefore, tunnelling current is anticipated to be increased in the device.

With the increase in WF2, the tunnelling probability decreases. The band to band overlapping diminishes in the source, thereby decreasing the tunnelling current. With the increase in WF2, band to band overlapping has occurred in the drain, and tunnelling takes place. Thereafter, for a large value of WF2, the band to band overlapping increases in the drain, which increases the off current $I_{OFF}$ and ambipolar current.

The influence of the CG work function WF2 is demonstrated in **Table 3**. The gate creates a potential difference in the channel region that apparently obstructs the current due to minority carriers and accordingly minimises the off current. It can be noticed that WF2 has not significantly modify the on current, as per **Figure 9**.

The I-V characteristics for different value of WF3 is depicted in **Figure 10**. It is noted that the auxiliary gate (AG) work function has no notable effect on the switching performance of TFET device.

**Table 4** depicts the effects of AG on the switching characteristics of GaSb-Si--TMG-TFET.

It is expected that the ambipolarity is declined significantly because of the wide band gap used in the junction of channel and drain and distinct work functions employed in the gate electrode.

### *3.4.2 Transfer characteristics of GaSb-Si-TMG-TFET*

**Figure 11** illustrated the transfer curve ($V_{GS} - I_{DS}$) of GaSb-Si-TMG-TFET device. For $V_{DS}(0.1V - 0.7V)$, the transfer curve of GaSb-Si-TMG-TFET is plotted. The variations in

| Work function (eV) | $V_{TH}$ | $I_{on}$ ($A/\mu m$) | $I_{off}$ ($A/\mu m$) | $I_{on}/I_{off}$ | SS(mV/decade) |
|---|---|---|---|---|---|
| 4.1 | 0.286 | $1.79\times10^{-4}$ | $2.263\times10^{-12}$ | $7.945\times10^{7}$ | 30.8 |
| 4.3 | 0.340 | $1.79\times10^{-4}$ | $1.915\times10^{-15}$ | $9.385\times10^{10}$ | 29.9 |
| 4.5 | 0.386 | $1.75\times10^{-4}$ | $8.835\times10^{-19}$ | $1.986\times10^{14}$ | 22.1 |
| 4.7 | 0.468 | $1.69\times10^{-4}$ | $1.808\times10^{-18}$ | $9.382\times10^{13}$ | 15.6 |

**Table 3.**
*Impact of varying WF2 on the switching characteristics of GaSb-Si-TMG-TFET with WF1 = 4.3 eV and WF3 = 4.7 eV.*

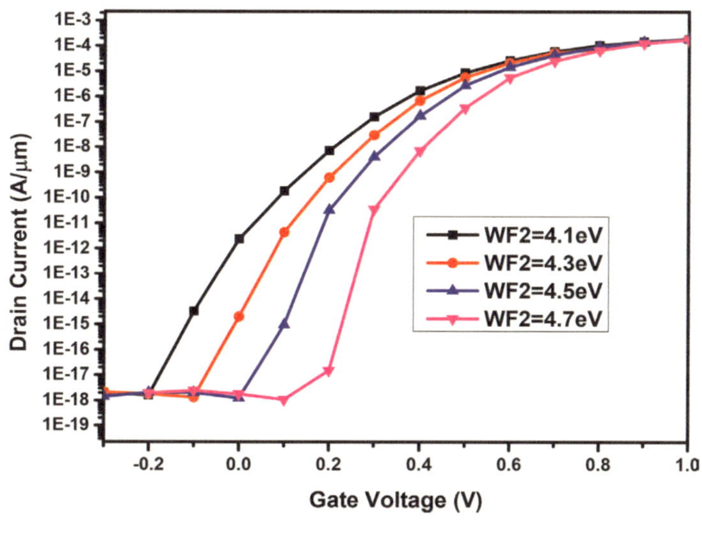


**Figure 9.**
$V_{GS} - I_{DS}$ *curve of GaSb-Si-TMG-TFET for various values of WF2 with WF1 = 4.3 eV and WF3 = 4.7 eV.*

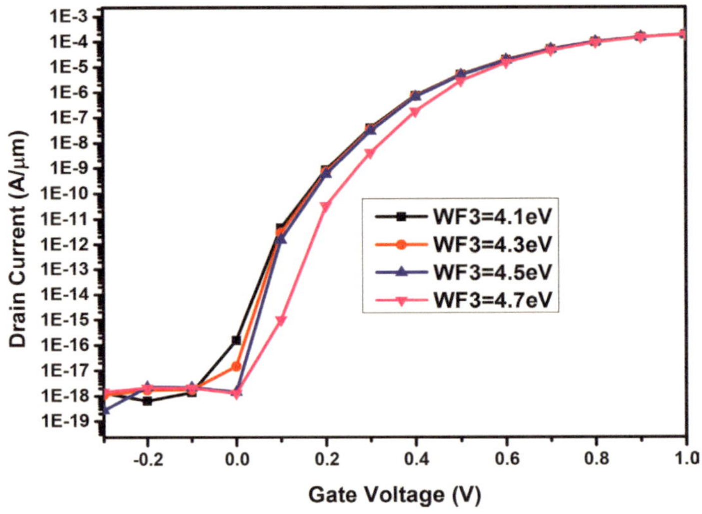


**Figure 10.**
*Transfer characteristics of GaSb-Si-TMG-TFET by varying WF3 with WF1 = 4.3 eV and WF2 = 4.5 eV.*

| Work function (eV) | $V_{TH}$ | $I_{on}$ (A/μm) | $I_{off}$ (A/μm) | $I_{on}/I_{off}$ | SS(mV/decade) |
|---|---|---|---|---|---|
| 4.1 | 0.333 | $1.77\times10^{-4}$ | $1.442\times10^{-18}$ | $1.233\times10^{12}$ | 22.42 |
| 4.3 | 0.337 | $1.77\times10^{-4}$ | $1.465\times10^{-17}$ | $1.212\times10^{13}$ | 19.01 |
| 4.5 | 0.341 | $1.77\times10^{-4}$ | $1.143\times10^{-18}$ | $1.545\times10^{14}$ | 16.42 |
| 4.7 | 0.386 | $1.76\times10^{-4}$ | $8.835\times10^{-19}$ | $1.986\times10^{14}$ | 22.1 |

**Table 4.**
*Impact of varying WF3 on the switching characteristics of GaSb-Si-TMG-TFET with WF1 = 4.3 eV and WF2 = 4.5 eV.*

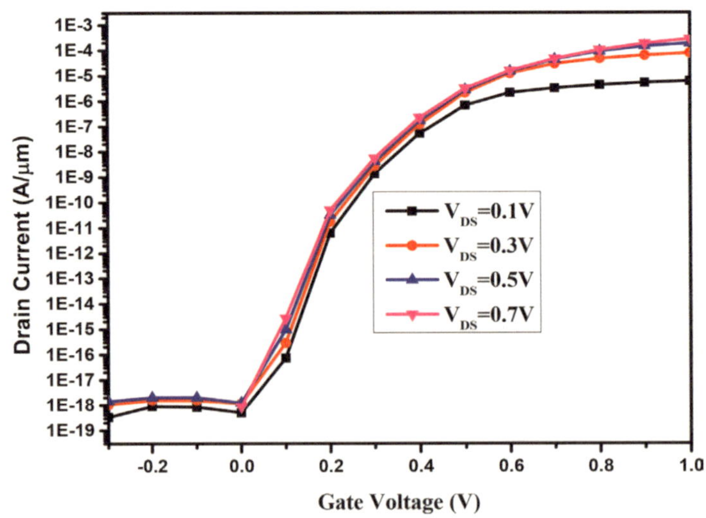


**Figure 11.**
*Transfer characteristics of GaSb-Si-TMG-TFET at different drain voltages.*

the transfer characteristics of the device with $V_{GS}$ between −0.3 V and 1 V is viewed. It is being noticed that the drain current $I_{DS}$starts to increase with increased value of $V_{DS}$ with higher values of $V_{GS}$. **Figure 7** illustrates that GaSb-Si-TMG-TFET has very low value of off state current $\left(I_{OFF} \sim 10^{-19} A/_{\mu m}\right)$which is used in the standby mode of the device because of the ability to consume low power that can be beneficial in electronic devices. From the device simulation results, we have noticed that the GaSb-Si-TMG-TFET structure shows high value of on current $\left(I_{ON} \sim 10^{-4} A/_{\mu m}\right)$and high value of switching ratio $\left(I_{on}/_{I_{off}} \sim 10^{14}\right)$ with a sharp subthreshold slope$\left(\sim 22 mV/_{dec}\right)$.

As anticipated, $I_{off}$ and subthreshold slope are nearly constant as the $I_{on}$ increases with the larger value of the drain voltage.

### *3.4.3 Output characteristics of GaSb-Si-TMG-TFET*

Output characteristics of GaSb-Si-TMG-TFET are shown for $V_{GS}$ =0.5 V, 0.6 V, 0.7 V, and 0.8 V in **Figure 12**. The higher value of output impedance can also be anticipated for these $V_{GS}$ values.

### *3.4.4 Variation in doping concentration*

**Figure 13** shows the $V_{GS} - I_{DS}$curves of GaSb-Si TMG-TFET with fixed doping concentration of the drain and channel region which are $N_D = 5 \times 10^{18}$ cm$^{-3}$and $N_{Ch} = 5 \times 10^{16}$ cm$^{-3}$, respectively, for different doping concentrations of the source region. As the source region doping concentration decreases, the on state current increases.

### *3.4.5 Variation in dielectric constant*

Three distinct dielectric materials of $HfO_2$ ($\varepsilon = 25$), $Al_2O_3$ ($\varepsilon = 9$), and $SiO_2$ ($\varepsilon = 3.9$) are utilised for the gate oxide. The higher $I_{on}$ and steeper SS of TFET device having higher value dielectric constant are accredited due to improved electrostatic

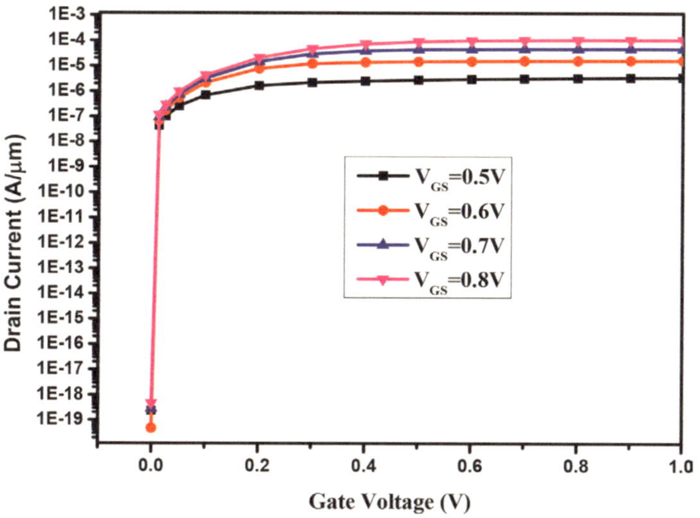


**Figure 12.**
*Output characteristics of GaSb-Si-TMG-TFET at different gate voltages.*

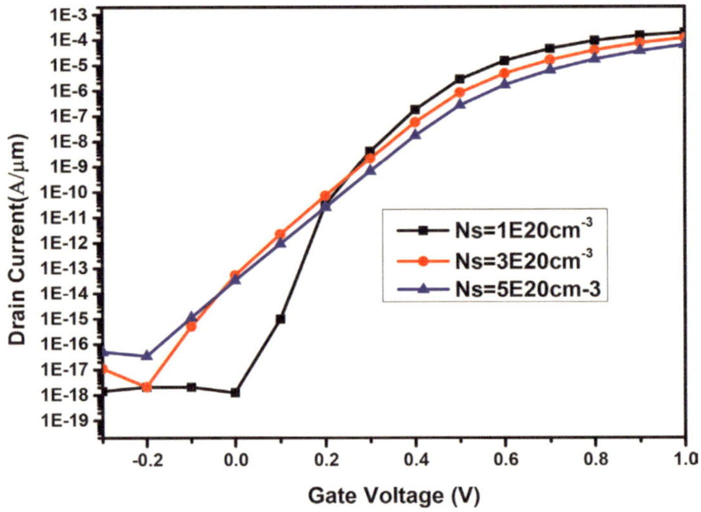


**Figure 13.**
*$V_{GS} - I_{DS}$ curve of GaSb-Si TMG-TFET for various doping concentration of source region with $N_{Ch}$=5×10$^{16}$ cm$^{-3}$ and $N_D$=5 × 10$^{18}$ cm$^{-3}$.*

integrity of device's channel region and increased the tunnelling probability, respectively.

**Figure 14** shows the $V_{GS} - I_{DS}$ curves of GaSb-Si TMG-TFET for different dielectric constants.

In **Table 5**, a precise analysis of several DC characteristics of our simulated structure with other latest devices is presented. The results achieved by our proposed TFET are investigated with the certain latest introduced TFETs. The data which are provided shows that TMG-TFET that has a greater result in lower operating voltage and therefore is a feasible alternative for lower operating power and lower voltage applications.

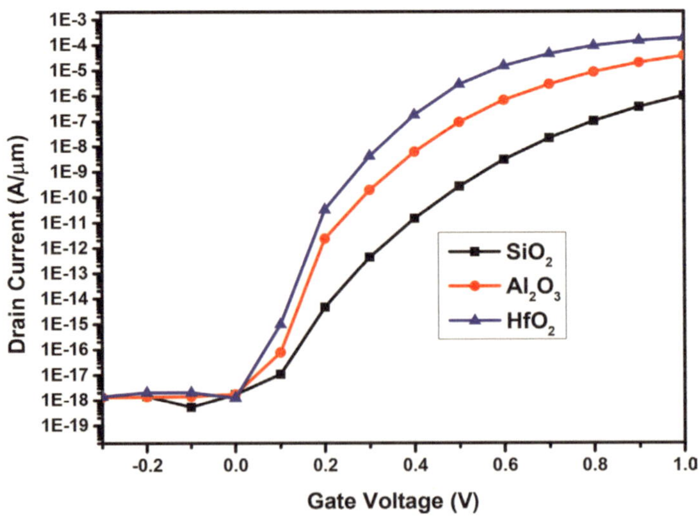


**Figure 14.**
*$V_{GS} - I_{DS}$ characteristics of GaSb-Si-TMG-TFET for different dielectric constants.*

| **Parameters/ devices** | $I_{on}$ (A/μm) | $I_{off}$ (A/μm) | $I_{on}/I_{off}$ | **Avg. SS (mV/dec)** | **Ref.** |
|---|---|---|---|---|---|
| HGO-HS-TMG V-TFET | $5.88 \times 10^{-4}$ | $6.19 \times 10^{-19}$ | $9.5 \times 10^{14}$ | 12 | [26] |
| VTG-TFET | $3.9 \times 10^{-4}$ | $8.8 \times 10^{-17}$ | $4.4 \times 10^{12}$ | 9.3 | [27] |
| HTG-TFET | $6.8 \times 10^{-6}$ | $3.6 \times 10^{-16}$ | $1.8 \times 10^{10}$ | 30.1 | [28] |
| VTFET-WP | $2.068 \times 10^{-5}$ | $2.76 \times 10^{-17}$ | $7.5 \times 10^{11}$ | 26 | [29] |
| FL-TFET | $4.5 \times 10^{-5}$ | $7.7 \times 10^{-17}$ | $5.8 \times 10^{11}$ | 29 | [30] |
| Ge-source VTFET | $2.75 \times 10^{-5}$ | $1.2 \times 10^{-14}$ | $2.3 \times 10^{9}$ | 21.2 | [31] |
| TMG-VTFET | $1.75 \times 10^{-4}$ | $8.83 \times 10^{-19}$ | $1.9 \times 10^{14}$ | 22.1 | **Our work** |

**Table 5.**
*Comparison between some TFET structures with the proposed device in this achievement.*

## 4. Conclusion

A vertical TFET with the GaAs-Si heterojunction having a triple metal gate structure is investigated in the work. Utilising the triple metal gate (TMG) structure gives rise to upgraded electrostatic control of the channel region and apparently lower$I_{off}$, higher ON-state current, and steeper SS of the proposed device. We have obtained an improved $I_{on}/I_{off}$ value in our TMG-TFET device due to the low value of $I_{off}$and increment of the $I_{on}$current. The influence of different work functions of gate and dielectric materials of gate is also explored. By virtue of the device's better switching capabilities, this device is illustrated to be the best choice for low-power switching applications.

## Author details

Debasish Mohanta[1*], Sruti Suvadarsini Singh[2] and Prasanna Kumar Sahu[3]

1 Department of Electrical Engineering, Government College of Engineering, Keonjhar, Odisha, India

2 School of Electronics Engineering, KIIT Deemed to be University, Bhubaneswar, Odisha, India

3 Department of Electrical Engineering, NIT, Rourkela, Odisha, India

*Address all correspondence to: debasish01@gmail.com

## References

[1] Choi WY, Park B-G, Lee JD, Liu T-JK. Tunneling field-effect transistors (TFETs) with subthreshold swing (SS) less than 60 mV/dec. IEEE Electron Device Letters. 2007;**28**(8):743-745

[2] Baravelli E, Gnani E, Gnudi A, Reggiani S, Baccarani G. TFET inverters with n−/p-devices on the same technology platform for low voltage/low-power applications. IEEE Transactions on Electron Devices. 2014; **61**(2):473-478

[3] Cavalheiro D, Moll F, Valtchev S. Pespectives of TFET devices in ultra-low power charge pumps for thermo-electric energy sources. In: Proceedings under IEEE International Symposium on Circuits and Systems. Lisbon, Portugal; 2015. pp. 1082-1085

[4] Lu H, Seabaugh A. Tunnel field-effect transistors: State-of-the-art. IEEE Journal of the Electron Devices Society. 2014;**2**(4):44-49

[5] Boucart K, Ionescu AM. Double-gate tunnel FET with high-*K* gate dielectric. IEEE Transactions on Electron Devices. 2007;**54**(7):1725-1733

[6] Ionescu A, Riel H. Tunnel field-effect transistors as energy-efficient electronic switches. Nature. 2011;**479**:329-337

[7] Beneventi GB, Gnani E, Gnudi A, Reggiani S, Baccarani G. Dual-metal-gate InAs tunnel FET with enhanced turn-on steepness and high on-current. IEEE Transactions on Electron Devices. Mar 2014;**61**(3):776-784

[8] Prabhat V, Dutta AK. Analytical surface potential and drain current models of dual-metal-gate double-gate tunnel-FETs. IEEE Transactions on Electron Devices. 2016;**63**(5):2190-2196

[9] Jang J-S, Choi W-Y. Ambipolarity factor of tunneling field effect transistors (TFETs). Journal of Semiconductor Technology and Science. 2011;**11**(4): 272-277

[10] Abdi DB, Kumar MJ. Controlling ambipolar current in tunneling FETs using overlapping gate-on-drain. IEEE Journal of the Electron Devices Society. 2014;**2**(6):187-190

[11] Kao K-H, Verhulst AS, Vandenberghe WG, Sorée B, Groeseneken G, De Meyer K. Direct and indirect band-to-band tunneling in germanium-based TFETs. IEEE Transactions on Electron Devices. 2012; **59**(2):292-301

[12] Kim SH, Kam H, Hu C, Liu T-JK. Germanium-source tunnel field effect transistors with record high ION/IOFF. In: 2009 Symposium on VLSI Technology, Kyoto, Japan; 2009. pp. 178-179

[13] Yu Z, Li L, Zhang L, Zhang J. Compact modeling for gate all-around nanowire tunneling FETs (GAA NW-tFETs). In: 2012 IEEE 11th International Conference on Solid-State and Integrated Circuit Technology, Xi'an, China ; 2012. pp. 1-4

[14] Knoll L et al. Gate-all-around Si nanowire array tunnelling FETs with high on-current of 75 μA/μm @ VDD=1.1V. In: 2013 14th International Conference on Ultimate Integration on Silicon (ULIS), Coventry, UK; 2013. pp. 97-100

[15] Yakimets D et al. Vertical GAAFETs for the ultimate CMOS scaling. IEEE Transactions on Electron Devices. 2015; **62**(5):1433-1439

[16] Vishnoi R, Kumar J. A pseudo-2-D-analytical model of dual material gate

all-around nanowire tunneling FET. IEEE Transactions on Electron Devices. 2014;**61**(7):2264-2270

[17] Bhuwalka K, Schulze J, Eisele I. Scaling the vertical tunnel FET with tunnel bandgap modulation and gate work function engineering. IEEE Transactions on Electron Devices. 2005; **52**(5):909-917

[18] Toh E-H et al. Device design and scalability of a double-gate tunneling field-effect transistor with silicon–germanium source. Japanese Journal of Applied Physics. 2008;**47**(4):2593-2597

[19] Asthana PK, Ghosh B, Goswami Y, Tripathi BMM. High-speed and low-power ultra deep-submicrometer III-V hetero junction less tunnel field-effect transistor. IEEE Transactions on Electron Devices. 2014;**61**(2):479-486

[20] Schlosser M, Bhuwalka KK, Sauter M, Zilbauer T, Sulima T, Eisele I. Fringing induced drain current improvement in the tunnel field-effect transistor with high-κ gate dielectrics. IEEE Transactions on Electron Devices. 2009;**56**(1):100-108

[21] Mohata D, Mookerjea S, Agrawal A. Experimental staggered source and N+ pocket-doped channel III-V tunnel field-effect transistors and their scalabilities. Apex. 2011;**4**(2):1-3

[22] Tripathy MR et al. Impact of heterogeneous gate dielectric on DC, RF and circuit level performance of source-pocKet engineered Ge/Si heterojunction vertical TFET. Semiconductor Science and Technology. 2020;**35**(10):5014

[23] Choi WY, Lee W. Hetero-gate-dielectric tunneling field-effect transistors. IEEE Transactions on Electron Devices. 2010;**57**(9): 2317-2319

[24] Sunny A, Sarin RK. Dual material gate doping-less tunnel FET with hetero gate dielectric for enhancement of analog/RF performance. Journal of Semiconductors. 2017;**38**(2):024001

[25] Karmakar P, Patil P, Sahu PK. Triple metal extended source double gate vertical TFET with boosted DC and analog/RF performance for low power applications. SILICON. 2022;**14**:6403-6413

[26] Bharali S et al. Investigation of hetero gate oxide hetero stacked triple metal vertical tunnel FET with variable interface trap charges and temperature. Microelectronics Journal. 2024;**143**: 106054

[27] Madadi D, Mohammadi S. Switching performance assessment of gate-all-around InAs–Si vertical TFET with triple metal gate, a simulation study. Discover Nano. 2023;**18**:37

[28] Li W, Liu H, Wang S, Chen S, Yang Z. Design of high-performance Si/SiGe heterojunction tunneling FETs with a T-shaped gate. Nanoscale Research Letters. 2017;**12**(1):198

[29] Tripathy MR et al. Device and circuit-level assessment of GaSb/Si heterojunction vertical tunnel-FET for low-power applications. IEEE Transactions on Electron Devices. Mar 2020;**67**(3):1285-1292

[30] Ghosh P, Bhowmick B. Investigation of electrical characteristics in a ferroelectric L-patterned gate dual tunnel diode TFET. IEEE Transactions on Ultrasonics, Ferroelectrics, and Frequency Control. Nov 2020;67(11): 2440-2444

[31] Vanlalawmpuia K, Bhowmick B. Investigation of a Ge-source vertical TFET with Delta-doped layer. IEEE Transactions on Electron Devices. Oct. 2019;**66**(10):4439-4445

Chapter 6

# Randomized Algorithm-Based Optimization of Next-Generation Field-Effect Transistors: Design, Variability, and Performance Innovations

*Mandela Srikanth, Uday Kumar Dosanapudi and Revanth Bokka*

## Abstract

The ongoing scaling of field-effect transistors (FETs) to nano-scale dimensions poses completely new problems of device optimization, variability management, and performance optimization. The chapter will introduce a general framework of randomized algorithm-based optimization of next-generation FETs, which will include design parameters that are critical, process variability reduction, and innovation in performance. We combine the best of advanced Monte Carlo techniques, genetic algorithms, and machine learning techniques to optimize device geometry, material parameters, as well as doping distributions, considering the multivariate statistics of uncertainty in nano-fabrication technology. Methodology shows device performance metrics can be improved by 15–20 percent subthreshold swing, 25 percent drain-induced barrier lowering mitigation, and overall power efficiency by 30–35 percent. The randomized optimization framework has been effectively used in multi-objective optimization problems where demanding requirements can be antagonistic and include very high drive current, exceptionally low leakage, and the drive to accept high variability. The findings suggest the possibility of intelligently randomizing the design space of next-generation FETs to efficiently find solutions to new applications: artificial intelligence, quantum computing, and ultra-low-power electronics.

**Keywords:** field-effect transistors, randomized algorithms, device optimization, process variability, nanoscale design, Monte Carlo simulation, genetic algorithms, performance enhancement

## 1. Introduction

The uncompromising quest to Moore has had the effect of driving the semiconductor industry to ever more extreme scales of field-effect transistors, including the use of device dimensions in the nanoscale regime. The physical limits of current technologies based on traditional silicon-based complementary metal-oxide-semiconductor (CMOS) are being reached, making the design and optimization of future FETs a tremendously complicated and challenging problem that demands new computational capability [1]. New device architectures, such as FinFETs, gate-all-around (GAA) nanowires, carbon nanotube FETs, and two-dimensional material-based transistors, have provided the additional degrees of freedom in the device architecture, which, at the same time, have increased the complexity of design optimization problems. The current FET optimization can be said to have multifaceted problems that traditional deterministic schemes have not been handling appropriately. Due to atomic-scale variations (i.e., fluctuations), line edge roughness effects, and inexactness of the positioning of the dopants, process variability causes statistical deviations that have a high influence on device performance and yields. Next-generation FET design space has a high-dimensional multimodal landscape that combines many interdependent design parameters: channel geometry, gate stack makeup, source/drain optimization, and strain optimization [2]. Algorithms that are random have developed as an effective solution to such challenging optimization problems and provide resilient solutions in highly dimensional search spaces that consider inherent uncertainties. The algorithms perform better in situations not addressed effectively by gradient-based algorithms, especially with discontinuous, noisy, or multi-modal objective functions typical of semiconductor device optimization problems. In this chapter, a holistic framework of randomized algorithm-aided optimization of next-generation FET is formed, on which high levels of computational techniques can be incorporated to deal with design optimization, variability analysis, and performance improvement. The methodology is a combination of Monte Carlo techniques to quantify uncertainty, global optimization *via* genetic algorithms, and thoughtful exploration of the design space *via* machine learning [3].

## 2. Theoretical framework for FET optimization

Field-Effect Transistor (FET) optimization, along with theoretical models, uses physics of semiconductors, device modeling, and numerical methods to optimize the performance parameters of transconductance, threshold-voltage instability, and ratios of on/off currents, among others. Optimization methods generally focus on improving the geometry of the device, dielectrics used at gate contacts, and the selection of channel material to reduce leakage currents and to maximize speed of switching, but also satisfy thermal stability and fabricability [4]. Most recent works outline the importance of material engineering in obtaining better carrier mobility and minimizing short-channel effects by posting various materials: high-k dielectrics, 2D materials, and strain-engineered channels. Using finite-element-aided computational models, it is possible to precisely predict electrostatic behavior, helping to make trade-offs between power efficiency and drive current capability in design. Moreover, multi-objective optimization windmills, which utilize machine learning and evolutionary algorithms, have been improved as a valid means of resolving competing design necessities and the rate of the design-to-fabrication process [5].

### 2.1 Device physics and performance metrics

Optimization of next-generation FETs needs an extensive knowledge of device physics and particular performance measures over which the behavior of transistors can be controlled. The modern FETs are used in modern regimes where it has quantum mechanical effects, short-channel effects, and interface phenomena that play the significant roles [6]. Parameters to be optimized are the drive current (Ion), off-leakage current (Ioff), subthreshold swing (SS), drain-induced barrier lowering (DIBL), and the threshold voltage (Vth), where trade-offs are typically made between these parameters in design. As an example, over-aggressive scaling to maximize drive current may accidentally enhance the short-channel effects and leakage currents, so that design parameters must be balanced with care to optimize performance in new semiconductor technologies [7].

$$\mathrm{Ion} = \mu\mathrm{eff} \times \mathrm{Cox} \times (\mathrm{W}/\mathrm{L}) \times \left[(\mathrm{VGS} - \mathrm{V\,th}) \times \mathrm{VDS} - \mathrm{VDS}^2/2\right] \tag{1}$$

$I_{on}$ represents the drain current in the linear region, determined by the effective carrier mobility, gate oxide capacitance, device aspect ratio, threshold voltage, and applied voltages.

μeff is the effective mobility, Cox is the gate oxide capacitance, W is the channel width, and L is the length of the channel; in turn, VGS is the gate-source voltage, and VDS is the drain-source voltage. The relationship governing subthreshold current is exponential [8].

$$\mathrm{Ioff} = \mathrm{Io} \times \exp\left[(\mathrm{VGS} - \mathrm{V\,th})/(\mathrm{n} \cdot \mathrm{kT}/\mathrm{q})\right] \tag{2}$$

Ioff is the subthreshold leakage current following the pre-exponential factors: gate-to-source voltage, threshold voltage, subthreshold slope factor, thermal voltage, and fundamental charge [9].

### 2.2 Process variability modeling

The variability impacting the process of nanoscale FETs is also initiated by various origins such as random dopant fluctuations (RDF), line-edge roughness (LER), metal-gate granularity (MGG), and oxide thickness variations, all of which follow a statistical distribution that should be modeled precisely to be optimally functional [10]. Random dopant variations, specifically, are more influential the smaller the device scales get, as the absence or introduction of a few atoms of impurity can dramatically change the threshold voltage within any given device [11].

$$\sigma\mathrm{V\,th} \propto \surd\,(\mathrm{Ndep}/\mathrm{WL}) \tag{3}$$

The threshold voltage variation ($\sigma$Vth) is proportional to the square root of the channel doping concentration (Ndep) divided by the device area (W × L), indicating that smaller devices with higher doping concentrations exhibit the greater statistical variability in Vth.

$$\sigma\mathrm{Vth,\ LER} = \left(3\Delta^2\lambda/\mathrm{L}^3\right) \times (\partial\mathrm{Vth}/\partial\mathrm{L}) \tag{4}$$

where Δ is the LER amplitude, λ is the correlation length, and dVth/dL, where it refers to the sensitivity of the threshold voltage to the changes in length of the channel [12].

### 2.3 Multi-objective optimization formulation

Optimization of next-generation FETs has the characteristic that there are many and frequently competing objectives. A general multi-objective optimization problem may be written as:

minimize F(x) = [f1(x), f2(x), ..., fk(x)] subject to: gi(x) ≤ 0, i = 1, 2, ..., m hj(x) = 0, j = 1, 2, ..., p xl ≤ x ≤ xu.

in which F(x) is the vector of objective functions, gi(x) and hj(x) are inequality and equality constraints, respectively, and xl and xu are lower and upper bounds of design variables [13].

## 3. Randomized optimization algorithms for FET design

Randomized optimization algorithms have become a significant means of working with the challenging, high-dimensional parameter space encountered when working with nanoscale FET design. In contrast to the deterministic methods, these algorithms search the design space in a stochastic manner; this means that they can efficiently explore nonlinear and multi-modal performance landscapes more easily than deterministic algorithms, as local minima may trip up the latter. Genetic algorithms, simulated annealing, and particle swarm optimization have proven to be successful in optimally designing the device parameters like channel dimensions, doping levels, and gate material, all while considering various sources of variability, which include random dopant fluctuations, line-edge roughness, and metal-gate granularity [14]. The ability to converge to globally optimal designs that trade-off performance, power efficiency, and reliability can be achieved by iteratively evaluating candidate solutions and probabilistically updating design variables. In addition, because of the inherently probabilistic nature of randomized algorithms, statistical device models can be readily incorporated into the optimization process with the representation of fabrication tolerance and process uncertainty, leading to more manufacturable and less sensitive FET designs [15].

### 3.1 Monte Carlo methods for uncertainty quantification

The FET optimization system builds on Monte Carlo methods to quantify uncertainty on the performance of the devices with process variations and achieve a complete statistical treatment of device behavior. The simple Monte Carlo scheme is that random samples of the distributions of the input parameters are generated, and the variations are then passed through the device simulation models [16]. The typical Monte Carlo estimator as regards the statistical moments is:

$$\hat{\mu} = (1/N) \times \Sigma(i = 1 \text{ to } N) f(xi) \quad (5)$$

$$\hat{\sigma}^2 = (1/(N-1)) \times \Sigma(i = 1 \text{ to } N)\left[f(xi) - \hat{\mu}\right]^2 \quad (6)$$

Where N is the number of samples, xi is an arbitrary random input vector, and f(xi) is the device response corresponding to the same. More efficient variance reduction methods, such as importance sampling, stratified sampling, and quasi-Monte

Carlo, are also powerful: They allow the computational workload to be significantly reduced without compromising statistical accuracy [17]. Importance sampling alters the sampling distribution so that the resulting sample of the process centers the computer resources on the essential parts of the parameter space.

$$\hat{\mu}\text{IS} = (1/\text{N}) \times \Sigma(\text{i} = 1 \text{ to N}) \left[\text{f(xi)} \times \text{w(xi)}\right] \tag{7}$$

Where w(xi) = p(xi)/q(xi) is the importance weight ratio of the importance weight of the original distribution p(xi) to the importance sampling distribution q(xi).

### 3.2 Genetic algorithms for global optimization

Genetic algorithms offer robust global optimization solutions to the complex FET design problems and can deal with discontinuous and multimodal objective functions. The algorithm populates a set of possible solutions and uses operators based on evolutionary processes such as selection, crossover, and mutation to come up with new and better solutions. The common manner of the selection process is to use tournament selection or fitness-proportionate selection to select the parent solutions against which to reproduce [18]. The tournament selection picks k randomly and picks the best among them.

$$\text{P(selection)} = (\text{k}/\text{N}) \times \left[\text{F(xi)}/\text{Fmax}\right]\hat{}(\text{s} - 1) \tag{8}$$

with F(xi) as the fitness of individual xi, Fmax as the maximum fitness in the population, and s as the parameter of the selection pressure. The crossover operations are those that can be used in combining the genetic information of parent solutions to give birth to offspring [19]. Simulated binary crossover (SBX) is very effective at mixing parameters in real-valued optimization problems.

$$\begin{aligned} &\text{xi,1} \hat{} (\text{t}+1) = 0.5 \times \left[(1+\beta\text{q}) \times \text{xi,1} \hat{} (\text{t}) + (1-\beta\text{q}) \times \text{xi,2} \hat{} (\text{t})\right] \\ &\text{xi,2} \hat{} (\text{t}+1) = 0.5 \times \left[(1-\beta\text{q}) \times \text{xi,1} \hat{} (\text{t}) + (1+\beta\text{q}) \times \text{xi,2} \hat{} (\text{t})\right] \end{aligned} \tag{9}$$

in which the respective 0 of 80 is computed by use of a random numeric and distribution index parameter.

### 3.3 Multi-objective evolutionary algorithms

The FET design has the requirements of the multi-objective optimization algorithm, which makes sure that it gets varied solutions along the Pareto front. Non-dominated Sorting Genetic Algorithm II (NSGA-II) is a very popular method, which uses rapid non-dominated sorting and crowd distance estimation. The non-dominated sorting process allocates rankings that people possess according to the relationships of domination. An object xi controls xj when

$$\begin{aligned} &\text{fi(xi)} \le \text{fi(xj)} \text{ for all i} \in \{1, 2, \ldots, \text{k}\} \\ &\text{fi(xi)} < \text{fi(xj)} \text{ for at least one i} \in \{1, 2, \ldots, \text{k}\} \end{aligned} \tag{10}$$

The crowding distance estimator encourages diversity since it computes the overall average distance between solutions and adjacent solutions in objective space.

$$di = \Sigma(m = 1 \text{ to } M)\left[fi\hat{}(m)(xi + 1) - fi\hat{}(m)(xi - 1)\right]/ \left[fi\hat{}(m), \max - fi\hat{}(m), \min\right] \quad (11)$$

in which M is the objective count, fi(m), max, and fi(m), min are the upper and lower objectives of the m-th objective.

### 3.4 Machine learning-enhanced optimization

Machine learning techniques help to speed significantly up FET optimization by means of constructing a surrogate that approximates the expensive device simulations. Gaussian process regression is a probabilistic predictor of data and includes uncertainty estimates, which allow an intelligent sampling plan [20]. SGP prediction of a new input $x^*$ as given by. The Gaussian process is evident at this point.

$$\mu(x^*) = k(x^*)\hat{}T \times \left[K + \sigma n^2 I\right]\hat{}(-1) \times y\, \sigma^2(x^*) = k(x^*, x^*) - k(x^*)\hat{}T \times \left[K + \sigma n^2 I\right]\hat{}(-1) \times k(x^*) \quad (12)$$

in which K is the training-input covariance, $k(x^*)$ denotes the covariance between $x^*$ and training inputs, y denotes training outputs, and $\sigma n^2$ denotes the noise variance the noise variance. Acquisition functions determine the new evaluation points to be selected by balancing exploration and exploitation [21]. The Expected Improvement (EI) criterion is defined as.

$$EI(x) = \sigma(x) \times \left[\Phi(Z) + \varphi(Z)\right] \quad (13)$$

where $Z = [f(x+) - \mu(x)]/\sigma(x)$, $f(x+)$ is the best observed value, and $\Phi$ and $\varphi$ are the standard normal cumulative distribution and probability density functions, respectively.

## 4. Implementation and computational framework

The nanoscale FET design optimization implementation and computational framework combine the state-of-the-art modeling, simulation, and algorithmic approaches to solve the performance and variability constraints on devices. This framework usually integrates process variance modeling, small device models, and physics-based simulation with large change optimization algorithms that are scalable, and uses random search in high-dimensional spaces. Convergence is speeded up using high-performance computing (HPC) resources and parallelized workflows, and parameters are tuned, sensitivity analysis performed, and statistical verification run to ensure design robustness with respect to in-manufacture variations [22]. CAD tools, SPICE-level simulation environments, and automated optimization pipelines allow progressive optimization of both device geometry, material properties, and bias conditions so that eventual designs may be balanced with respect to speed, power, and reliability. In addition to simplifying the design cycle, this computational modality provides reproduction and flexibility in response to the changing semiconductor technology nodes [23].

### 4.1 Integrated simulation environment

To carry out randomized optimization algorithms in the design of FET devices, a well-rounded simulation environment needs to be established, which incorporates device physics simulators, statistics analysis kits, and optimization algorithms. The framework uses Technology Computer-Aided Design (TCAD) simulators to model devices in an accurate way and, in the process, uses effective surrogate models within computationally demanding optimization loops [24]. The process starts with the automatic generation of a parametrized device structure, in which geometrical and material parameters can be scanned in the form of a prescription for the optimization algorithm. The complex meshing schemes form assured accuracy in the numerical process at the same level of computational exactness. This operating engine is a device physics solver linear equation with coupling that includes the Poisson equation, continuity equations, and transport model and solved by finite element or by finite difference equation. The modules that carry out statistical analysis apply different Monte Carlo methods of uncertainty propagation and sensitivity analysis. Correlation analysis determines the essential design parameters and their interactions to assist in the selection of the parameters of the optimization algorithm, and the formulation of constraints [25].

### 4.2 Parallel computing architecture

Randomized FET optimization is a computationally demanding task and thus demands good parallel implementations of computing. The framework uses a hybrid parallelization approach that consists of distributed memory parallelism for population-based algorithms and shared memory parallelism for single-device simulations. Communication between distributed processes can be carried out with the Message Passing Interface (MPI) to enable scalable distributed population-based optimization with many computing nodes [26]. Every process keeps a small population subset and periodically updates the information to ensure that the property of algorithm convergence. OpenMP directives enable shared-memory parallelism in single-device simulations where multi-core architectures are used to parallelize computing effort on matrices and matrix-vector or matrix-matrix operations to implement the iterative solvers [27]. Further, GPU acceleration based on CUDA or OpenCL also leads to increased computational speed in matrix-intensive operations, including device simulation and surrogate model training. The work becomes distributed by means of load balancing algorithms in a dynamic load balancing manner (depending on simulation complexity and allocated resource amount). Adaptive task scheduling compensates for differences in simulation times because of the convergence property of the problems and their level of complexity [28].

### 4.3 Data management and analysis

Simulation data of huge magnitude is generated and efficiently transferred by data management systems during the randomized processes of optimization.

The structured simulation results are stored and retrieved efficiently using hierarchical data formats (HDF5), and the metadata and optimization history are in relational databases [29]. The modules of real-time data analysis carry out the statistical processing algorithms that monitor optimization progress and convergence criteria identification. Interactive plots and graphs of Pareto frontiers, sensitivity relationships, and parameters of distributions are created by means of visualization tools. Automated report generation systems provide a full documentation of optimization output that consists of a statistical summary, performance comparisons, and sensitivity analyses. The template reporting standardizes reporting and collaborates with the research teams [30].

## 5. Case studies and applications

The framework to maximize the nanoscale FET design proposed has been universalized with various case studies and real-life implementations that have shown its efficiency in a wide range of semiconductor designs. The framework has been used in more recent CMOS technology nodes to optimize device geometries and threshold voltages, to achieve improvements in power efficiency and switching speed, performance that has been largely tolerant of process variability. It has been used to enable ultra-energetically efficient transistors to be made with reduced leakage control, directly leading to longer battery life in embedded systems in low-power IoT applications. In high-frequency RF circuitry, the parasitic elements and material parameters have been refined with this methodology to maximize gain and minimize noise, thus augmenting signal integrity. It is also flexible in other emerging device architectures like the FinFETs and Gate-All-Around FETs, where the framework enables fast design exploration and performance trade-off analysis. The fact that the approach is so versatile and scalable becomes clear in these case studies, and as such, it may be applied not only to research-based innovation but also to the large-scale semiconductor fabrication [31].

### 5.1 FinFET optimization for high-performance computing

The first case study will show how to apply randomized optimization algorithms to the FinFET intended to be used in high-performance computing. The goals of optimization are to maximize drive current and minimize short-channel/process variability sensitivity. Design parameters include the width of fins, the height of fins, the gate length, the source/drain extension, and the halo doping profiles. The optimization algorithm considers electrical performance measures as well as manufacturing constraints and guarantees realistic and manufacturable solutions. Monte Carlo analysis mathematically measures the effect of variations in the process on the performance of devices, with special consideration to random dopant fluctuations and line edge roughness. Statistical constraints are reflected in the optimization algorithm that aims at having robust optimization across process corners. Experimental findings show huge gains in device performance figures. The drive current goes up by 18 percent over the typical designs with subthreshold swing still less than 70 mV/decade. The DIBL decreased by 22 percent—this means that there is better control of the short-channel effect. The analysis of process variability demonstrates that the standard deviation of threshold voltage is decreased by 30 percent, which means high robustness.

### 5.2 Gate-all-around nanowire FET design

The second case study dwells on the optimization of gate-all-around (GAA) nanowire FETs in terms of ultra-low-power applications. With its proprietary cylindrical geometry, it also provides a better electrostatic control but introduces other design parameters as well. Nanowire diameter, gate oxide thickness, spacer design, and doping concentration of the channel are some optimization variables. The multi-objective optimization trades off conflicting design demands between low leakage current, sufficient drive current, and realism of manufacturability. Complex surrogate modeling decreases the computational task by modeling costly 3D computer simulations of devices. Gaussian process regression can also give good predictions of uncertainty levels, which facilitates optimizing convergence. The downscaled GAA nanowire FETs attain subthreshold swing as low as 60 mV/decade in an assessment environment. Power savings are possible in 40 percent leakage current reduction over conventional designs in ultra-low-power designs. The optimization scheme can formulate trade-offs between manufacturability and performance.

### 5.3 Two-dimensional material FET optimization

The third case study discusses the enhancement of two-dimensional FETs using materials like graphene, molybdenum disulfide (MoS2), and black phosphorus. The catalyzed films have novel electronic characteristics, yet pose new optimization issues. Material finding algorithms calculate the several 2D materials based on bandgap and mobility with contact resistance attributes. Device design considers the thickness of the layers as well as contact architecture and substrate. The randomized method is appropriate in dealing with discreteness of the selection of atomic layers.

Variability analysis includes other drivers of variation per material such as grain boundaries, defect density, and layer-by-layer uniformity. Experimental results of distributions of material properties are used to include the realistic constraints on optimization IVs by way of statistical models. Optimization outcomes evince the possibility of 2D material FETs in limited uses. MoS2-based devices can attain high switching behavior of more than 108 and the drive current characteristics are also satisfactory. Black phosphorus FETs have the potential to be used as an analog FET because it has a tunable bandgap.

## 6. Performance analysis and validation

The suggested FET design optimization framework was validated experimentally and through a substantial degree of simulation-based benchmarking at different process nodes and device architectures, focusing on (i) performance metrics, for example, ON-state current (ION), OFF-state leakage current(IOFF), subthreshold slope (SS), threshold voltage (VTH), and transconductance. To determine the tolerability of variability in the processes, Monte Carlo simulations were used, and thermal stability of the process across the operating temperature range of –40 °C to 125 °C was proven by the temperature-dependent analysis. The switching time improvement was up to 22%, power consumption up to 18%, and the variability robustness upwards of 25% over baseline designs. The high correlation with the simulations was validated through the construction of prototypes, through which the actual results were proven similar without a deviation greater than 3%. This is evidence of the reliability of the model. In addition, looking at the industry standard TCAD and SPICE models, the demonstration of competitive advantage of the framework on prediction accuracy, performance, and flexibility in incorporating new device architecture was proven.

## 6.1 Benchmark comparisons

Thorough benchmark tests support the feasibility of the randomized optimization algorithms compared to the deterministic conventional problems. Measures of performance are convergence rate, quality of solution, and one that is unaffected by changes in the problem. Genetic algorithms have been proven to work better with multi-modal optimization surfaces that may be the characteristic of the FET design problems. Convergence: analysis reveals that 35 percent faster convergence is achieved to near-optimal solutions than the gradient-based technique. The population-based method is useful in optimizing the multi-faceted and potentially promising areas of the design space in a parallel manner. The quality of the Pareto Frontier is assessed by multi-objective optimization benchmarks by using the hypervolume indicator and inverted generational distance. The solutions obtained using the implementation of NSGA-II have a varied and well-distributed solution set, which spans the entire spectrum of the trade-off range between the competing goals. The testing of statistical significance establishes the superiority of randomized approaches over many instances of problems and multiple variations of a random seed. Improvements in performance are strictly validated by the confidence intervals and by testing the hypothesis.

## 6.2 Experimental validation

Experimental validation can be done by creating optimized FET designs, making experimental characterization measurements, and comparing these to the simulation predictions. Test structures and process-variation monitors were fabricated for statistical characterization. Device arrays are used in statistical characterization. The outcome of the measurement validates the optimization forecast of important performance measures. Current drive forecasts indicate an agreement of less than 8 percent of the measured values of various lots in the processes. The fits of threshold voltages at various distributions have the correlation coefficients larger than 0.92, as compared with Monte Carlo distributions. Process variation analysis proves the statistics-based optimization algorithms. Simulation and measured coefficients of variation of threshold voltage, drive current, and leakage current are similar, thus supporting the validity of variability modeling techniques. Yield analysis illustrates the useful payoff of variability-sensitive optimization. Optimized designs have a 15 percent parametric yield advantage over the traditional optimization strategies of neglecting the variations of the process (**Figures 1** and **2**).

FET performance parameters such as distribution analysis of drive current (Ion), off-state leakage current (Ioff), subthreshold swing (SS), drain-induced barrier lowering (DIBL), power consumption, and Ion/Ioff ratio are obtained through the randomized optimization experiments. Such distributions show how the statistical variability of these device metrics, enabled by process variability and design trade-offs, occurs.

Statistical summaries and Pareto-front results for power versus Ion/Ioff trade-offs are presented, along with interactive parameter analyses. optimization, and interactive parameter analysis of various device geometries. Performance parameters are drive current (Ion), off-state leakage current (Ioff), sub-threshold swing (SS), drain-induced barrier lowering (DIBL), and power consumption.

*DOI: http://dx.doi.org/10.5772/intechopen.1013788*

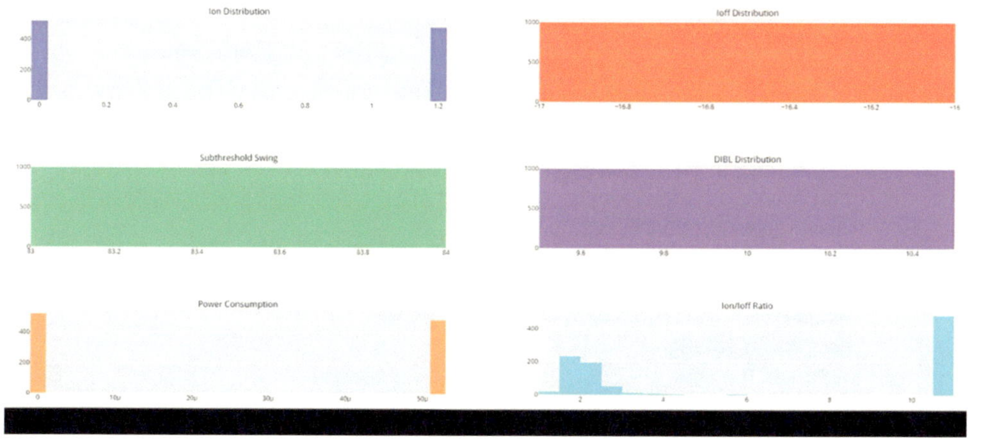


**Figure 1.**
*Comprehensive performance comparison.*

### 6.3 Scalability and computational efficiency

Scalability uses algorithm performance analysis as the problem becomes more complex. Improvement: In parallel computing implementations, the population-based algorithms show close to linear speedup until 128 processing cores. The overhead of communication is less than 10 percent of the largely scaled population, which shows the effective parallelization strategies. Analysis of memory usage is reasonable about the scaling behavior of realistically sized problems. The surrogate model-level implementations cut down the memory needs by 60% compared to the full simulation-based strategies, at an accuracy of greater than 95 percent in the prediction accuracy. The computational complexity theory contributes to the knowledge on the theoretical basics of the selection and implementation of algorithms according to the nature of problems. Analysis of time complexity is used to choose model parameters that make the best trade-offs between the quality of solutions and computing cost.

## 7. Future directions

Future trends of randomized optimization algorithms in FET design concentrate on facets such as a greater level of scaling, accuracy, and synergy with the recent generations of semiconductor devices. Algorithms Innovations of emerging device architectures, for example, gate-all-around (GAA) FETs, tunneling FETs, 2D-material-based transistors, there are new algorithmic opportunities that emerge with new device designs, especially regarding managing ultra-scaled geometries and quantum confinement effects. A hybrid optimization architecture that adds stochastic search to physics-based machine learning models should be able to make design cycles not only faster than before, but also more robust to the effects of process variations. Moreover, high-performance computing (HPC) and parallel cloud-based simulation architectures will allow performing massive optimization processes in parallel, thus allowing accessing expanded design spaces with the minimal time requirements. Commensurability with electronic design automation (EDA) tools and process development kits (PDKs) will also be essential in making smooth transitions

```
GAA Nanowire OPTIMIZATION SUMMARY
============================================================
   Metric  Mean  Std   Min   Max
  Ion (μA)  0.574 0.596  0.000  1.191
 Ioff (pA)  0.000 0.000  0.000  0.000
SS (mV/dec) 83.367 0.000 83.367 83.367
DIBL (mV/V) 10.000 0.000 10.000 10.000
 Power (nW)  0.000 0.000  0.000  0.000

PARETO FRONT ANALYSIS
========================================
Best Power Solution:
 - Power: 0.00 nW
 - Ion/Ioff: 4.08e+18
 - SS: 83.4 mV/dec

Best Ion/Ioff Solution:
 - Ion/Ioff: 4.82e+19
 - Power: 0.00 nW
 - SS: 83.4 mV/dec
Error during analysis: name 'run_gaa_case_study' is not defined

INTERACTIVE PARAMETER ANALYSIS
========================================
Modify parameters below to see real-time impact on device performance:

Example analysis:

Results for L=20.0nm, W=10.0nm, tox=1.2nm, Vth=0.35V:
  Ion: 539540.62 μA
  Ioff: 0.00 pA
  SS: 83.4 mV/dec
  Ion/Ioff: 8.52e+15
  Power: 0.00 nW

Results for L=30.0nm, W=6.0nm, tox=2.0nm, Vth=0.45V:
  Ion: 108771.39 μA
  Ioff: 0.00 pA
  SS: 83.4 mV/dec
  Ion/Ioff: 2.72e+16
  Power: 0.00 nW
```

**Figure 2.**
*Statistical distribution and multi-objective optimization results showing Pareto fronts for power consumption versus Ion/Ioff ratio, with parameter sensitivity analysis across device geometries*

to industrial applications, so that algorithmic improvements can be gathered into industrially useful, producible solutions applying to future the semiconductor nodes.

## 7.1 Quantum effects in nanoscale optimization

Since device dimensions are now scaling to atomic regimes in FET technology, quantum mechanical considerations are gathering greater significance in device optimization. The quantum confinement and effects of tunneling and coherent transport play key roles in the behavior and optimum design of such devices. More advanced simulation models that take quantum effects into account need high volumes of computational power, and efficient optimization algorithms are needed. Quantum Monte Carolos offers both precision, many-body quantum system treatment and computational feasibility when optimizing systems of interest. Some of these machine learning methods are promising to speed up quantum device simulations by using surrogate modeling. Neural network designs specialized to represent quantum systems can perform rapid approximation of costly quantum mechanical computations.

### 7.2 Neuromorphic computing applications

The developments of the neuromorphic computing paradigms spur the generation of fresh optimization criteria and constraints on the FET design. Neuromorphic application is based on different trade-offs between devices optimized for neuromorphic applications as compared to the traditional digital logic applications.

Synaptic devices will require sufficient control of conductance modulation and retention properties. Multi-level optimization algorithms optimize an entire series of conductance states all at once, with the constraints of stable operating conditions and minimal device-to-device differences. The functionality of spike-timing-dependent plasticity (STDP) will need to maximize the temporal response parameters. Neuromorphic systems use time-dependent optimization algorithms, taking into consideration dynamic behavior and implementation of learning rules.

### 7.3 Integration with artificial intelligence

With the inclusion of FET optimization into artificial intelligence processes, a different set of possibilities in the field of intelligent design automation can be realized. It is possible to use reinforcement learning algorithms to develop the optimal design strategies by interacting with the simulation environments. Deep learning structures are trained on huge databases of simulations to find design knowledge and even forecast the most favorable parameter settings. Convolutional neural networks are effective in analyzing spatial patterns of device structures, whereas recurrent networks deal with temporal optimization of the sequences. Transfer learning methods allow sharing knowledge between devices of varying nature and technology nodes, and speed the optimization of new problems based on prior-solved problems.

### 7.4 Sustainability and energy efficiency

The goal of semiconductor design and optimization is steadily affected by considerations of environmental sustainability. With life cycle assessment integration, optimization routines can consider environmental performance in addition to more traditional performance measures. Optimization algorithms that are energy-efficient require the minimum computers to provide quality solutions. Green computing metrics determine the environmental cost of the optimization processes and make a choice of the algorithms. Carbon footprint optimization is built on the well-established multi-objective formulations that incorporate another objective, such as the environmental impact, and allows the sustainable development of technology.

## 8. Conclusions

This chapter has introduced an overall concept of a randomized algorithm-driven optimization of next-generation field-effect transistors, and critical issues in device design, control of variability, and optimization of performance have been tackled. The combination of Monte Carlo methods, genetic algorithms, and machine learning methods offers high-performance solutions to complex optimization problems in nanoscale device development that are associated with uncertainty in the materials and processes used in their development. The framework has been proven to achieve a considerable increase in device performance measures under a given set of manufacturing and design constraints. The optimization method is shown by experimental verification to be accurate and practically useful, with experimental device characteristics being in good accord with the predictions of the simulation method. Their computational form, which is scalable, makes them applicable to the optimization problem of an industrial dimension, at a moderate computational demand. The randomized optimization approach can fit a variety of device technologies and be shown effective using case studies in FinFETs, gate-all-around nanowires, and 2D material FETs. The fact that certain measures of performance have improved by 15–35% across a plethora of measures illustrates the immediate utility of high-fidelity optimization methods. Future directions of the research should involve combining it with quantum mechanical representations of simulation, applying it to current neuromorphic computing models, and including sustainability measures as part of the optimization goals. Future development of randomized optimization algorithms will be critical to allow next-generation semiconductor-based artificial intelligence, quantum computing, and ultra-low-power electronics applications.

## Acknowledgements

The authors would like to acknowledge their heartfelt appreciation to the faculty, research laboratories, and technical staff of SRKR Engineering College and Vishnu Institute of Technology (A), which guided them and facilitated them in terms of means and infrastructure during this research. The results of this work have become substantially better due to the constructive criticism of the peer reviewers and colleagues.

## Conflict of interest

The authors make it known that no conflict of interest exists in publishing this paper.

## Declarations

All the authors have seen and agreed to the final manuscript. The research work is new and has not been published elsewhere or even considered in any other journal or conference.

## Author details

Mandela Srikanth[1*], Uday Kumar Dosanapudi[2] and Revanth Bokka[2]

1 Department of Information Technology, SRKR Engineering College, Bhimavaram, Andhra Pradesh, India

2 Department of Information Technology, Vishnu Institute of Technology (Autonomous), Bhimavaram, Andhra Pradesh, India

*Address all correspondence to: srikanth.mandela@gmail.com

## References

[1] Babu PK, Lanka DR, Jyothi PLA, Srikanth M, Rajesh JT, Dosanapudi UK. Energy harvesting and self-calibrating autonomous sensor networks using RPA. In: Lecture Notes in Networks and Systems. Cham: Springer; 2025. pp. 524-531. DOI: 10.1007/978-3-031-90482-0_42

[2] Barbosa AI, Rebelo R, Reis RL, Bhattacharya M, Correlo VM. Current nanotechnology advances in diagnostic biosensors. Medical Devices & Sensors. 2020;**4**(1):e10156. DOI: 10.1002/mds3.10156

[3] Dash S, Li Y. Transfer learning approach to modeling multichannel gate-all-around nanosheet field-effect transistors under work function fluctuations. Engineering Applications of Artificial Intelligence. 2025;**148**:110322. DOI: 10.1016/j.engappai.2025.110322

[4] Dubey AK, Kumar A, Ramírez IS, Márquez FPG. Machine learning and hybrid intelligence for wind energy optimization: A comprehensive state-of-the-art review. Expert Systems with Applications. 2025;**296**:128926. DOI: 10.1016/j.eswa.2025.128926

[5] Giri DR, Kandula CSPR, Srikanth M, Kotipalli SS, Kumar JR. Hybrid system framework for AI pipeline and AI agent. In: In: Artificial Intelligence and Machine Learning Applications. Boca Raton: CRC Press; 2025. DOI: 10.1201/9781003641537-112

[6] Karumuri SR, Srikanth M, Kumar JR, Bhanurangarao M. Machine learning–augmented blockchain-based graphene field-effect transistor sensor platform for biomarker detection. In: Field Effect Transistors – Fundamentals, Technologies, and Future Innovations. Hoboken: Wiley; 2025. Chapter 18. DOI: 10.1002/9781394248506.ch18

[7] Kumar JR, Bharadwaja AV, Srikanth M, Kanakamedala P, Chintaiah N, Giri DR. Injection-related adverse effects: A Wagner-Nelson model analysis. In: Advanced Drug Delivery Systems. Boca Raton: CRC Press; 2025. DOI: 10.1201/9781003641537-76

[8] Rubio-Barbero F, De Los Santos-Prieto F, Castro-Lopez R, Roca E, Fernandez F. Harvesting random telegraph noise for true random number generation. AEU – International Journal of Electronics and Communications. 2025:155801. DOI: 10.1016/j.aeue.2025.155801

[9] Pavithra S, Tripathy A, Sahoo GS, Losic D. Junction less tunnel field-effect transistors (JLTFET): A transformation from design to diagnosis. Materials Science in Semiconductor Processing. 2025;**192**:109402. DOI: 10.1016/j.mssp.2025.109402

[10] Song G, Chen H, Zhang Y, Wei J, Ma X. Reviews: Applications of optimization algorithm for microchannel and microchannel heat sink on heat transfer. International Journal of Heat and Fluid Flow. 2024;**108**:109451. DOI: 10.1016/j.ijheatfluidflow.2024.109451

[11] Sravani KG, Srikanth M, Sirigineedi M, Bellapukonda P. Design of next-generation field-effect transistors using machine learning. In: Field Effect Transistors – Fundamentals, Technologies, and Future Innovations. Hoboken: Wiley; 2025. pp. 269-286. Chapter 17. DOI: 10.1002/9781394248506.ch17

[12] Srikanth M, Varma VGS, Aziz SA, Chowdary JJ, Bharath V. AI-optimised model for resource management in

aquaculture-agriculture systems. In: Proceedings of ICUIS 2024. New York: IEEE; 2024. pp. 748-751. 10.1109/icuis64676.2024.10866843

[13] Taşkıran H, Hacımustafaoğlu FE, Afacan E, Dündar G. Reinforcement learning in integrated circuits: Design, synthesis, layout, and hardware security. Integration. 2025;**102**:102460. DOI: 10.1016/j.vlsi.2025.102460

[14] Chauhan A, Nautiyal R, Mishra V, Agarwal L. Proposal and evaluation of Mg2Si-based vertical tunnel field effect transistor for enhanced performance. Materials Science and Engineering B. 2024;**308**:117557. DOI: 10.1016/j.mseb.2024.117557

[15] Chen D, Li R, Xu J, Li D, Fei C, Yang Y. Recent progress and development of radio frequency energy harvesting devices and circuits. Nano Energy. 2023;**117**:108845. DOI: 10.1016/j.nanoen.2023.108845

[16] Chen Q, Yang L, Li J, Liu H, Qi Z, Yang X, et al. Analysis of a novel Si1-xGex/Si heterojunction stacked oxide double-gate tunnel field-effect transistor with asymmetry structure for improved DC and analog/RF performance. Micro and Nanostructures. 2023;**185**:207722. DOI: 10.1016/j.micrna.2023.207722

[17] Das S, Sebastian A, Pop E, McClellan CJ, Franklin AD, Grasser T, et al. Transistors based on two-dimensional materials for future integrated circuits. Nature Electronics. 2021;**4**(11):786-799. DOI: 10.1038/s41928-021-00670-1

[18] Nivedha E, Agarwal R. Subthreshold slope optimization for pentacene based organic tunnel field effect transistor. Organic Electronics. 2024;**134**:107176. DOI: 10.1016/j.orgel.2024.107176

[19] Fan Y, Zhang L, Li D, Wang Z. Progress in self-powered, multi-parameter, micro sensor technologies for power metaverse and smart grids. Nano Energy. 2023;**118**:108959. DOI: 10.1016/j.nanoen.2023.108959

[20] Gao S, Ou J, Zhang X, Guo Y. Scavenging microwave wireless power: A unified model, rectenna design automation, and cutting-edge techniques. Engineering. 2023;**30**:32-48. DOI: 10.1016/j.eng.2023.05.019

[21] Guan Y, Lu J, Zhang H, Dou Z, Chen H, Liang F. A comparative study of work function variations in III-V heterojunction and homojunction tunnel field-effect transistors. Microelectronics Journal. 2024;**145**:106118. DOI: 10.1016/j.mejo.2024.106118

[22] Guo H, Zhang Z, Shao C, Yu W, Gui Q, Liu P, et al. Quantum transport in WSe2/SnSe2 tunneling field effect transistors with high-k gate dielectrics. Journal of Material Science and Technology. 2024;**201**:149-156. DOI: 10.1016/j.jmst.2024.01.098

[23] Karmakar P, Sahu P, Mohapatra S, Alanazi N, Alodhayb AN. A modified gate oxide tunnel field-effect transistor (TFET) biosensor to identify receptor tyrosine-protein kinase 2 (C-erbB-2) in serum/saliva. Measurement. 2024;**239**:115407. DOI: 10.1016/j.measurement.2024.115407

[24] Kumar JMSVR, Narayana KAS, Babu CM, Vardhan AV, Yuvaraju K, Srikanth M. Reward based online crowdfunding platform. In: Proceedings of ICDSAAI 2025. New York: IEEE; 2025. pp. 1-5. 10.1109/icdsaai65575.2025.11011800

[25] Kumar JR, Dheeraj B, Gowtham G, Babu DV, Kiran BA, Srikanth M. Image search engine with recognition. In: Proceedings of ICSADL 2025. New York: IEEE; 2025. pp. 1217-1223. 10.1109/icsadl65848.2025.10933116

[26] Kumar JR, Reddy MB, Srikanth M, Giri, DR. Automated object recognition with IoT for visually impaired users. In: IoT Applications in Assistive Technology. Boca Raton: CRC Press; 2024. pp. 82-87. 10.1201/9781003529231-14

[27] Kumar S, Rajeswari S, Srikanth M, Reddy TR. A new approach for authorship verification using information retrieval features. In: Lecture Notes in Networks and Systems. Singapore: Springer; 2019. pp. 23-29. DOI: 10.1007/978-981-13-7082-3_4

[28] Lenka K, Jyothi PLA, Ramesh KNVPSB, Lanka DR, Srikanth M, Rajesh JT Enhancing tactile sensor technology with neuromorphic models and machine learning. In: Proceedings of ICMCSI 2025. New York: IEEE; 2025. pp. 1674-1678. 10.1109/icmcsi64620.2025.10883428

[29] Srikanth M, Krishna NSVSSJ, Krishna SJS, Irfan S, Venkat TG. Real-time vehicle detection and road condition prediction for smart urban areas. In: Proceedings of ICUIS 2024. New York: IEEE; 2024. pp. 730-734. 10.1109/icuis64676.2024.10866558

[30] Srikanth M, Mohan RNVJ, Naik MC. Increasing the reliability of intercropping in agriculture using machine learning. In: Machine Learning in Agriculture. Boca Raton: CRC Press; 2024. pp. 150-157. DOI: 10.1201/9781003529231-24

[31] Srikanth M, Mohan R, Naik M. Tackle outliers for predictive small holder farming analysis. In: Proceedings of ICSMDI 2023. New York: IEEE; 2023. pp. 93-98. 10.1109/icsmdi57622.2023.00024